THE MINUTE BOOKS OF
THE SUFFOLK HUMANE SOCIETY

A painting in the Lowestoft and East Suffolk Maritime Museum at Sparrows Nest, Lowestoft, showing the Frances Ann setting off on 22 October 1820 to the Woodbridge sloop Sarah & Caroline on the Newcome Sand. She saved the crew of five from the sloop and also took off seven men from the brig George of London, which sank soon afterwards. There are several versions of this picture, one of which was used to produce a print. Reproduced by courtesy of Lowestoft and East Suffolk Maritime Museum.

THE MINUTE BOOKS OF
THE SUFFOLK HUMANE SOCIETY

A PIONEER LIFESAVING ORGANISATION AND THE WORLD'S FIRST SAILING LIFEBOAT

1806–1892

Edited by

ROBERT MALSTER

General Editor

DAVID BUTCHER

The Boydell Press

Suffolk Records Society
VOLUME LVI

A Suffolk Records Society publication
First published 2013
The Boydell Press, Woodbridge

ISBN 978–1–84383–805–0

Issued to subscribing members for the year 2012–2013

The Boydell Press is an imprint of Boydell & Brewer Ltd
PO Box 9, Woodbridge, Suffolk IP12 3DF, UK
and of Boydell & Brewer Inc.
668 Mt Hope Avenue, Rochester, NY 14620, USA
website: www.boydellandbrewer.com

The publisher has no responsibility for the continued existence or
accuracy of URLs for external or third-party internet websites referred to
in this book, and does not guarantee that any content
on such websites is, or will remain, accurate or appropriate

A catalogue record for this book is available
from the British Library

Papers used by Boydell & Brewer Ltd are natural, recyclable products
made from wood grown in sustainable forests

Printed and bound in Great Britain by
CPI Group (UK) Ltd, Croydon, CR0 4YY

CONTENTS

ILLUSTRATIONS

In memory of my old friends
Hugh D.W. Lees, Jack Mitchley and Jack Rose,
Lowestoft historians.

PREFACE

Although Lowestoft did not gain its harbour until 1831 the town had been involved in fishing, maritime trade and shipbuilding long before that. And, like those of any east coast town, its inhabitants saw many wrecks both on the shore and on off-lying sandbanks such as the Holm, which lies just off the town, and the Newcome a little to the south.

The story of the great storm of 18 December 1770, told by Edmund Gillingwater, the Lowestoft historian, is often quoted. The wind, which had been blowing from the south-west, abruptly backed to north-west and for two hours 'raged with a fury that was hardly ever equalled'.

> Anchors and cables proved too feeble a security for the ships, which instantly parting from them, and running on board each other, produced a confusion neither to be described nor conceived. At daylight a scene of the most tragic distress was exhibited; those who first beheld it assert that no less than eighteen ships were on the sand before this place at one and the same time, and many others were seen to sink; of those on the sand, one half were entirely demolished, with their crews, before nine o'clock; the rest were preserved a few hours longer; but this dreadful pause served only to aggravate the destruction of the unhappy men who belonged to them, who betook themselves to the masts and rigging; these continuing breaking, eight or ten were not unfrequently seen to perish at a time, without the possibility of being assisted.
>
> Fifteen only, about two in the afternoon, were taken off one of the wrecks, and about as many more were saved by taking to their boats, or getting on board other ships when they boarded each other. It is impossible to collect with certainty how many lives, or how many ships were lost in this terrible hurricane ...[1]

While many of the seamen drowned quickly, for others the agony was long and cruel. Those on shore watched the drama unfold, quite unable to give any aid to the unfortunate men facing death little more than half a mile away. They saw the masts of one ship go overboard; two men who had taken refuge in the rigging struggled through the water and over the wreckage and managed to climb on to the hull of the wreck. That afternoon a pilot boat put off from Lowestoft beach and went to the wreck, but it could not get near enough to take the men off.

The next day, to the astonishment of everyone ashore, it was seen that one of the men was still alive. About midday some of the beachmen launched a boat and got near enough to call to the man, but the breakers on the sand and the presence of floating wreckage that threatened to smash into their boat made it impossible to bring the boat alongside the ship. The beachmen shouted to him to clamber the length of the deck and to drop into the boat from the bows as the boat was run close to the ship, but he called that he was too weak to make his way forward as they suggested.

---

[1] Eyewitness account by Robert Reeve, a Lowestoft merchant and lawyer, quoted in Gillingwater 1790, pp. 65–6.

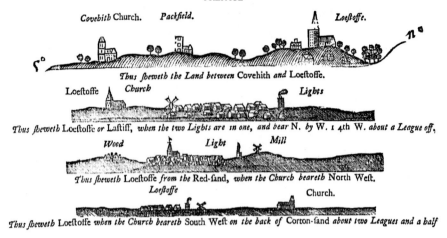

1. The coast in the vicinity of Lowestoft as seen from the sea. An illustration from John Seller's The Coasting Pilot for Great Britain and Ireland *(London, 1796). Reproduced by courtesy of Cambridge University Library.*

'The ensuing night put a period to his misfortunes and his life', Gillingwater's account concludes and he recommends that the inhabitants of Lowestoft should consider 'the possibility of constructing a vessel, or some other machine, upon such principles as should not be liable to overset, but should be capable of approaching any vessel in distress, during the most violent storms, and when surrounded with the most tumultuous waves ... The pleasure of saving so many valuable lives, will be ample recompence for any expence or trouble that may attend its execution.'[2]

Lowestoft gained its first lifeboat in 1801, just before a similar boat was sent to the village of Bawdsey, some two miles north of the entrance to the River Deben. The high hopes entertained by its promoters were not realised, for that boat was not favoured by the local seamen and they adamantly refused to use it. In 1807, however, the presence in the town of a London coachmaker, Lionel Lukin, a man who had earlier experimented with what he called an unimmergible (unsinkable) boat, led to the introduction of a different type of lifeboat, built by one of the local boatbuilders. That boat, the world's first sailing lifeboat, designed by Lukin on the lines of the yawls employed by the beachmen on salvage and other work, very soon came under the control of the Suffolk Humane Society, which had been set up the previous year.

There were already lifeboats stationed at the mouth of the Tyne and on the estuary of the Mersey before these boats were acquired for use in Suffolk, but the involvement of the Suffolk Humane Society and the introduction of a sailing lifeboat brought an entirely new dimension to lifesaving on the British coasts. The *Frances Ann*, as the new boat was named, after the daughter of Lord Rous, president of the Suffolk Humane Society, remained in service for forty years and in that

---

[2] Edmund Gillingwater, *An Historical Account of the Ancient Town of Lowestoft* (London, 1790), pp.65–67.

time is said to have saved more than three hundred lives, according to the Duke of Northumberland's report of 1851. She also provided a model for many of the later lifeboats that carried out much good work on the East Anglian coast.

The Humane Society did not merge with the Suffolk Association for Saving the Lives of Shipwrecked Seamen when that organisation came into being in 1824, as might have been expected, but continued to operate the lifeboats at Lowestoft and Pakefield until it handed over to the Royal National Lifeboat Institution in 1873, though working in association with the national institution from 1855. Its seniority among lifesaving organisations and its long period of operation, together with its having brought into service the first sailing lifeboat, makes its story of national if not international significance.

The editor first tried to trace the Humane Society's records when working on the history of East Anglia's lifesaving services in about 1970. A contemporary newspaper account of the society's final meeting revealed that the minute books and other records had been placed in 'a tin box' and deposited at the Lowestoft Sailors' and Fishermen's Home in Commercial Road, Lowestoft. Inquiries at the Bethel on Battery Green, the successor to that organisation, drew a blank, and it was said at that time that the records had been buried for safe keeping during the Second World War and had been ruined by damp.

It was some thirty years later that the editor was speaking on the telephone to a friend at Lowestoft who blandly said 'I've been reading the minute book of the Suffolk Humane Society.' It was quickly established that this was in the care of the Lowestoft Fishermen's Widows' and Orphans' Fund (now the Lowestoft Fishermen's and Seafarers' Benevolent Society), and in due course the secretary of that organisation, Mr Hugh Sims, very kindly made it possible for the surviving books to be transcribed for publication.

ACKNOWLEDGEMENTS

Many people have contributed to uncovering the history of the Suffolk Humane Society, a pioneering lifesaving association that operated the Lowestoft and Pakefield lifeboats over a period of more than half a century before it handed over to the Royal National Lifeboat Institution. First and foremost came Jack Mitchley and Eric Porter of the Port of Lowestoft Research Society, whose work did so much to reveal the beginnings of the lifeboat service in Suffolk. Sadly they are no longer with us to see the publication of this volume. Neither is Trevor Westgate, who alerted the editor to the existence of the early minute book, which had been thought to have been destroyed at the end of the Second World War.

The editor is most grateful to the Lowestoft Fishermen's and Seafarers' Benevolent Society and to its secretary Hugh Sims for the loan of the minute books. Dr E.E. Cockayne kindly provided information on William Henchman Crowfoot and commented on the case of Sergeant Bubb which led to the formation of the Suffolk Humane Society. Dr Alastair Massey of the National Army Museum helped to trace Sergeant Bubb's background and Gillian Hutchinson of the National Maritime Museum identified the ship wrecked at Kessingland and provided vital clues to the circumstances of her stranding on the Suffolk coast. Colin Dixon and James Woodrow gave invaluable assistance with illustrations, and the Port of Lowestoft Research Society and its treasurer, Stuart Jones, provided generous grants towards the cost of printing.

Thanks are also due to my general editor, David Butcher, for overseeing the transcription of the minute books, and to the co-ordinating editor of the Suffolk Records Society, David Sherlock.

GLOSSARY

air cases: wooden boxes covered with canvas shaped to fit into a lifeboat to provide buoyancy if or when the boat is flooded. Early boats such as the *Frances Ann* were equipped with sealed empty barrels lashed within the boat to provide the necessary buoyancy, but these were superseded by air cases in later boats.

beach company: associations formed by members of the maritime community in coastal towns and villages to carry on salvaging and similar work. They would own at least one yawl and probably other boats in which to carry on that work.

beachman: a man who participated in the work of a beach company, putting pilots aboard ships, attending on vessels in the roads and carrying out salvage operations.

billyboy: a type of coasting vessel, mainly from the Humber, Yorkshire and Lincolnshire, double-ended, flat-bottomed and having leeboards, and rigged as a sloop, ketch or schooner. Many were registered at Hull, and the name is said to be derived from the nickname given to inhabitants of the town of Kingston-upon-Hull, renowned for their adherence to Dutch William; alternatively it might be derived from the medieval bilander, a vessel that sailed *by the land* – a coaster.

brig: a two-masted sailing vessel, square rigged on both masts.

bumpkin: a fitting projecting ahead of the stem to which was attached the clew of the foresail.

coxswain: the man in charge of a boat, particularly a lifeboat. In the early days of the Suffolk Humane Society, Lieutenant Samuel Carter was known as commander of the lifeboat, but was in effect the coxswain, a term which had not then come into general use in connection with lifeboats.

creepers: a grapnel.

foresail: the sail set on the foremast. In the case of yawls and the early lifeboats there were usually three masts, fore, main and mizzen, the mainmast often being left ashore in heavy weather. When it became customary for large boats to carry only two masts east-coast seamen referred to the foremost one as the foremast and the other as the mizzen, and this nomenclature was of course applied also to the sails.

gat, gatway: a narrow passage between sandbanks, derived from Scandinavian languages.

gig: a fine-lined pulling (rowing) boat used by the beach companies in fine weather to put pilots aboard ships or to go ahead of a yawl to secure a job. Rowing six or eight oars, they were very fast craft.

irons: wrought-iron fittings to hold oars, spars, etc.

lugsail: a fore-and-aft sail extended at the head by a yard such as was employed in the yawls and lifeboats and also in fishing craft.

mizzen: the aftermost sail of a boat set on the mizzen mast.

outligger: a lengthy spar extending beyond the stern of a yawl or lifeboat to which the sheet of the mizzen sail was led, always so pronounced by Suffolk beachmen.

pulk-hole: a small pond or drainage sump.

punt: a small inshore fishing boat launched from the beach.

roads: the roadstead between the shore and the offlying sands which provided a sheltered anchorage for shipping.

schooner: a sailing vessel with two or more masts, with fore-and-aft sails on both or all masts.

sloop: a sailing vessel with a single mast and a fore-and-aft mainsail and headsails.

smack: a fishing boat, particularly one used for lining or for trawling; a passage boat, particularly one running a regular service between Leith, on the Firth of Forth near Edinburgh, and London; sometimes a term used as an alternative to sloop.

thole-pins, thowles: wooden pegs fitted into the gunwale of a boat in pairs to accommodate the loom of an oar.

thwart: a transverse wooden seat in a boat on which the rowers sit; supported by knees, the thwart also forms part of the structure of the boat, tying the the sides of the boat together.

yawl: a fast lug-rigged open boat, usually of 45–60 feet in length, employed by the beach companies. Pronounced yoll.

EDITORIAL METHODS

The minutes have been transcribed just as they were set down by the various secretaries, sometimes in idiosyncratic styles. In order to retain the flavour of the documents the variations of style have been reproduced largely as they are in the original. There are occasionally gaps in the minutes when, presumably, the secretary's memory failed him. Sometimes the secretaries made fairly obvious errors, one repeating three successive items when writing the minutes and another apparently writing the date 1846 when he should have written 1845. Misspellings of names such as *Stamford channel* for Stanford have been corrected in footnotes and the identities of committee members, etc., have also been amplified in footnotes.

Sums of money are, of course, rendered in pounds, shillings and pence and no attempt has been made to render these in a decimal conversion; changing values would in any case render any such attempt worthless. It might seem that a half-crown, two shillings and sixpence, was an inadequate reward for a man who leapt into the sea to save a life, but that sum was in the early nineteenth century by no means as insignificant as it appears today.

In the old monetary system there were twelve pence in one shilling and twenty shillings to the pound. The normal style used in the minutes is £1 2. 6 for one pound, two shillings and sixpence, 5/- for five shillings and 5/6 for five shillings and sixpence, but secretaries were not consistent in their style. In one instance a man was reported to have been rewarded with S5 for rescuing another from drowning; it is assumed that this is five shillings, but if so it is a most unusual variant. £1 1s. was a guinea.

The page numbers of the minute books are inserted in **bold** type within square brackets.

Transcripts of news items and letters published in local newspapers contemporary with the events described have been used where possible to fill the gap left by the loss of the missing minute book and also to augment the often terse terms of the minutes. It is felt appropriate to use these since in some cases at least they were written by the secretary of the society, and in other cases they appear to be excerpts from the minutes. Some particularly lively accounts of lifeboat services were penned by the Revd Bartholomew Ritson, curate of Lowestoft, during the lengthy period that he was secretary.

NOTES ON THE DOCUMENTS

Only two minute books of the Suffolk Humane Society survive. The status of the older of them, a board-covered leather-spined book some 12½ inches by 8 inches, is somewhat puzzling, since although the minutes run only from 31 July 1833 to 25 May 1859 the early pages contain a full account of the events leading up to the formation of the Society and of the earliest meetings of subscribers. A pencilled note, possibly made by Colonel George Seppings, a member of the committee in 1856, states that 'Thus far this appears to be a duplicate of the 1st. Minute Book – Here follows a continuation of the Minutes from the 1st. Vol commencing 31 July 1833 to present time Geo. S. Oct 1 56'. One assumes that he had the first minute book, now missing, to compare.

The label on the front cover bears the words 'Suffolk Humane Society Minute Book' and the date 1806, but somebody has added in ink 'Vol. 2'. One is left wondering if this might have been the original minute book, used in the early days of the Society and then for some reason discarded, to be brought back into use in 1833 as a second volume. There is some evidence to support this theory.

At the back of the book a single page bears a list of donations and annual subscriptions entered upside down. This is undated, but from the names entered it would seem to be from the early days of the Suffolk Humane Society and it would appear that it had been intended to enter subscription income each year at the back of the minute book; perhaps after this entry had been made it was decided to obtain a separate book for this purpose.

This, together with the label on the front cover, suggests that this might be an original minute book which for some reason ceased to be used after 9 September 1806. It was agreed at a meeting on 5 February 1806 that minute books should be procured for use by the secretaries, of whom there were two, the Revd William Spurdens and the Revd Michael Maurice. It is clear from the minutes of the meeting that each of the secretaries should have his own book. It might be that each secretary kept his own set of minutes at the beginning, and that it was then decided that this was unnecessary, leading to one of the books becoming disused, to be brought back into use when the other was full.

The minutes are at times rather carelessly written, and from time to time there are gaps where, perhaps, the secretary has been unable to decipher a name in his notes. In one case the heading to a meeting in 1846 reads 'A meeting of the [blank] held at the Queens Head Inn ...' and in another place the minutes remain incomplete, coming to an abrupt end in the middle of a sentence. Elsewhere the secretary has duplicated items within the minutes of a meeting. One would have expected such errors to be corrected when the minutes were read at the next meeting, but it might well be that this was not the normal procedure; many minutes are not signed by the chairman; the early ones are signed by one or other, or both, of the secretaries.

The other book is smaller, 8 inches by 6¼ inches, bound in similar fashion to the one just described and bearing on its front cover a label bearing the words

'Suffolk Humane Society July 1890'. It records the endeavours of George Edwards[1] to recover the remaining funds of the Society, seventeen years after it had handed over its lifeboat establishments to the Royal National Lifeboat Institution, and to put them to good use. Inserted on the first page, held down by pieces of gummed stamp paper, is a sheet of paper bearing a note that R.H. Reeve was treasurer to the Suffolk Humane Society when he died on 18 October 1888 and that 'on application to his executors for the books and papers relating to the society, some of them only were to be found, the third Vol. of the minute books (supposed to be the last) being still missing has necessitated the providing The Present Volume 1890'. The last entry in this book is dated 29 March 1892.

---

[1] George Edwards was resident engineer to the Norwich and Lowestoft Navigation Company and was involved with the construction of Lowestoft harbour. He was also responsible for the building in 1847 of the tied-arch bowstring bridge of cast iron at St Olave's, Herringfleet, a classic example of its type. In later life he was a magistrate and a man of some influence in the Lowestoft area. For an illustration of the bridge and a description of its design, said to be unique in East Anglia, see E.A. Labrum, *Civil Engineering Heritage: Eastern and Central England* (Thomas Telford for the Institution of Civil Engineers, 1994), pp. 129–30.

INTRODUCTION

The Suffolk Humane Society

The Suffolk Humane Society was formed at a meeting at Kessingland in 1806 after the pattern of the Royal Humane Society of London. Established as a result of a successful act of resuscitation performed by a surgeon from Beccles, Mr William Henchman Crowfoot, in December the previous year, the Society was intended to reward those who saved their fellows from drowning and those like Mr Crowfoot who resuscitated people who had apparently died. It took on additional duties the following year when it undertook the financing and administration of the Lowestoft lifeboat *Frances Ann*, the world's first sailing lifeboat and one of the most successful early lifesaving craft in Britain.

The story of the wreck of a troopship on Kessingland beach on 18 November 1805 is told in a letter from William Henchman Crowfoot (Plate 2) to the treasurer of the Royal Humane Society, Dr Hawes, which is reproduced in the minute book (pp.3–4 below). Crowfoot had been born at Kessingland in 1780 into a family that produced several generations of surgeons, was apprenticed to his uncle, William Crowfoot, when he was fourteen, and became his uncle's partner when he was twenty-five. He practised in Beccles for most of his life, dying in 1848 of typhus after performing a post-mortem on a typhus victim.

Crowfoot had been visiting a patient in Kessingland on the morning following the wreck. After spending an uncomfortable and indeed perilous night on the ship's deck the officers and 138 soldiers of the 28th Foot (later under Cardwell's reforms to become the Gloucestershire Regiment) had been brought ashore by local fishermen and were at the time being formed up and marched off to Lowestoft.

As he went down the road past the semi-ruinous St Edmund's Church he saw coming towards him a cart loaded with the belongings of the officers and in the charge of a red-coated sergeant of the 28th. Following it up the road was another cart, accompanied by four local farmworkers who had themselves been down to the beach earlier. Crowfoot spoke to the sergeant, who told him that the cart behind contained the body of another sergeant of the regiment who had collapsed on deck the previous night; thinking that the man was drunk, the officers left him there all night with the waves washing over him and other men treading on what they thought to be his dead body as they busied themselves about the ship. When the soldiers and the ship's crew were taken off that morning they had lowered the body by ropes into one of the boats, and it had been lying on the beach ever since until the four farm labourers took pity and fetched a farm cart to take the body up to the village.

Tiny Spark of Life

Crowfoot went across to the second cart and took a look at the body. He quickly realised that there was no smell of drink about the body; the man had collapsed not from drunkenness but from exposure and exhaustion; what would today be termed

2. William Henchman Crowfoot, 1780–1848, the Beccles surgeon whose resuscitation of a soldier given up for dead led to the formation of the Suffolk Humane Society.
Reproduced by courtesy of Beccles and District Museum.

hypothermia. And a close examination revealed a tiny spark of life remaining. The surgeon told the farm labourers to get him to an inn as quickly as they could, and there he set to work, aided by the four 'peasants', John Holmes, John Durrant, Joseph Durrant, and Charles Chipperfield.

> It was now about noon; and the man had been supposed dead thirteen hours [Crow-foot said in his letter to Dr Hawes]. I had him stripped, wiped dry, and put in warm blankets: and after three or four hours unremitted perseverance in the means which I judged most likely to restore suspended respiration, I had the inexpressible satisfaction of finding I had succeeded in rescuing a fellow-creature from premature death and in preserving to his country a robust young soldier.

Sergeant James Bubb, the soldier whose life Crowfoot saved, went on to serve his country for a number of years, eventually being discharged from the army at the age of forty-seven. Like many other men of the 28th, also known parenthetically at that time as the North Gloucestershire Regiment and in army slang as *The Old Braggs*, Bubb was a native of Gloucestershire, and it is very likely that he had fought in the Battle of Alexandria in 1801 when the regiment won the distinction of being allowed to wear a badge at the back of the shako as well as in the normal position. During the battle, which took place while an expeditionary force was driving a French army out of Egypt, the 28th found themselves attacked from the rear by French cavalry. Undaunted, the commanding officer gave the command '28th, front rank stand fast, rear rank about face'; the attack was repulsed. The 28th also fought with Wellington's army throughout the Peninsular War, 1808–14; presumably Sergeant Bubb was with his regiment and took a full part in the campaigns in which it fought.

The *Peace*, the ship that came ashore at Kessingland, was one of a fleet of transports that had left the Downs for the River Weser with the intention of landing British troops in Oldenburg. They encountered a very severe storm off the Dutch coast and the ships became scattered, most of them being forced to turn back; some, including the *Derwent*, found refuge in Harwich harbour.[1] The sending of 26,000 men to the Weser at that time was part of a plan to counter Napoleon's expansion in Europe by striking into northern Germany, occupying Hanover and marching into Holland and Flanders and on to Paris. That plan was negated by the refusal of Frederic William of Prussia to co-operate.[2]

The resuscitation of Sergeant Bubb, for which Crowfoot was awarded a silver medal by the Royal Humane Society (Plate 3), was a triumph in itself, but it was also to have effects that would outlive those involved in the event. At a meeting held, appropriately, at the King's Head in Kessingland on 7 January the following year it was decided to form a local society on the lines of the Royal Humane Society to promote 'whatever means may most effectually awaken the exertions of humanity in cases of shipwreck'. That was indeed the inaugural meeting of the organisation, but it was only at a meeting a month later at the Queen's Head Inn at Lowestoft, which housed the Lowestoft assembly rooms, that it was agreed 'that this Society shall be distinguished by the name of the Suffolk Humane Society'. The Queen's Head, in Tyler's Lane (now Compass Street), was in later years a regular meeting place of the Humane Society.

Throughout its existence the Society concerned itself with making rewards to those who saved others from drowning: half a guinea each to three men who rescued three others from a boat that upset in a heavy surf in the North Roads near the Battery, four shillings to William Crow for saving the life of a five-year-old boy who fell into Robert Tripp's pulk-hole (a good East Anglian word for what others might call a pond),[3] half a crown to another man who waded into the sea up to his waist

[1] *Ipswich Journal* 21 December, 1805.
[2] J. Steven Watson, *The Oxford History of England: The Reign of George III, 1760–1815* (1960), pp.424–25.
[3] There were several of these pulk-holes on the Denes, holding water that had drained down from the town's built-up area above the cliff. They must have proved both an attraction and a grave danger to children.

*3. The Royal Humane Society silver medal awarded to William Henchman Crowfoot,
which is now on display at Beccles and District Museum. Photograph courtesy of
James Woodrow.*

and brought out a child who had been 'taken off by the draw of the sea'. Yet right
from the beginning it sought to play a more proactive role in lifesaving.

It was at a third meeting in Beccles on 3 March 1806 that a suggestion was made
that the Society might unite with the fund raised by Robert Sparrow, of Worlingham
Hall and Sparrow's Nest, Lowestoft, for providing a lifeboat at Lowestoft. The
outcome of that suggestion was important for the Society and for the development
of lifesaving services not only in Suffolk but in a far wider field.

The Society not only controlled the Lowestoft lifeboat from about 1807 but from
1840 onwards was also responsible for the one stationed at neighbouring Pakefield.
By the time the first county-wide lifesaving association was formed in Norfolk in
1823 the Suffolk Humane Society had been operating very successfully for more
than sixteen years, showing the way to later organisations. The Norfolk Shipwreck
Association (its short popular title) was followed in 1824 by the Suffolk Association
for Saving the Lives of Shipwrecked Seamen, at the same time that the National
Institution for the Preservation of Life from Shipwreck was established; the name
was later changed to the Royal National Lifeboat Institution. A county organisation
was set up in Lincolnshire in 1827 and in the following year an Anglesey Association for the Preservation of Life from Shipwreck was inaugurated, unusually by a
lady, Frances Williams, on lines somewhat similar to the Suffolk Humane Society.[4]

---

4   See A. James, 'Frances Williams and the Anglesey Association for the Preservation of Life from
Shipwreck, 1827–1857', *Anglesey Antiquarian Society and Field Club Transactions* (1957), p.20.

The First Lifeboats

The story of the Lowestoft lifeboats goes back to 1800, when Robert Sparrow joined forces with the Revd Francis Bowness, the incumbent of Gunton, a coastal parish immediately to the north of Lowestoft, to raise funds for the building of a boat designed by Henry Greathead, of North Shields. Dr Hawes contributed five guineas (£5.25) to the fund on behalf of the Royal Humane Society.

The son of the supervisor of salt duties at South Shields, Henry Greathead went to sea at the age of about twenty but became a boatbuilder in 1785.[5] He built his first lifeboat in 1789 for use at the mouth of the Tyne. Like the other boats that followed from Greathead's boatyard, it was a pulling boat, equipped only for rowing, and was intended for lifesaving work within a fairly short distance of the shore. The Lowestoft boat was the sixth lifeboat built by Greathead, and was followed within weeks by a seventh boat for Bawdsey (Plate 4), about two miles north of the mouth of the Deben.

Ipswich Journal 4 October 1800

LIFE BOAT

The accounts of the very great Benefits that have arisen in the North parts of England, from the use of Life Boats, in saving lives of many wreck'd and distressed mariners, have encouraged some Gentlemen to set on foot a subscription for the purpose of building one for the Suffolk coast, of a similar construction, to be called the Lowestoft Life Boat, and to be stationed at Lowestoft; although it is meant that its usefulness should be extended to the North and South of that place, as far as may be consistently with safety.

Such persons as are willing to promote this benevolent scheme are requested to send their subscriptions to Robert Sparrow, Esq., of Worlingham-Hall, near Beccles, or the Rev. Francis Bowness, at Lowestoft.

SUBSCRIBERS

	£ s		£ s
Robert Sparrow, Esq.	5 5	Lieutenant Col. Harvey	1 1
George Dorset, Esq.	3 3	Thos. Boddington, Esq.	1 1
Rev. B. Sparrow, Clk	3 3	Rich. G. Townley	1 1
Rev. F. Bowness, Clk	3 3	John Elph, merchant	1 1
Rev. S. Kilderbee, Clk	2 2	Samuel Chambers, ditto	1 1
Lieutenant King	1 1	Thomas Russell	1 1
Ald. Chas Arnold, Esq.	3 3	Dr. Hawes, for the	
James Averard, Esq.	2 2	Humane Society	5 5
Rev. R. Potter, V	1 1	G.L. Reed, Esq.	5 5
Marquis Townsend	2 2	Chas. Berners, Esq.	2 2
Marchioness Townsend	2 2	Chas. Berners jun.	2 2
Major Gen. Loftus	2 2	Lord Bishop of Norwich	5 5
Robt. Mason, Pilot	1 1	Mrs. King	1 1
Dowager Countess		Lord Rous	2 2
Albemarle	2 2	Lady Rous	2 2
Dowager Countess		James Hatch, Esq.	3 3
Dysart	2 2	J.R. Abdy	1 1

---

[5] Adrian Osler, *Mr. Greathead's Lifeboats* (Newcastle upon Tyne: Tyne and Wear Museums Service, 1990), p.20.

Mason Cornish	1 1	James Hatch jun.	1 1
Rev. M. Maurice	1 1	Hon. F. Rous	1 1
Robert Reeve	2 2	Henniker Major Esq.	5 5
John Kittridge	1 1	Mrs. Dorset	2 2
Thomas Hunt	1 1	Thomas Manners	
Charles Pearse	1 1	Sutton, Esq.	2 2

LIFE BOAT

Before we can expect to receive the Credit and Assistance we solicit, it is necessary to detail to the public some account of Life Boats, and of the plan we have undertaken. The first Life Boat was established at South Shields, about 8 years ago, and was found so very beneficial, that another has been since provided by the liberality of the Duke of Northumberland, and endowed by him with a stipend of about 90£ per ann. Sunderland, Whitby, and Scarborough, have lately applied for models, and it is supposed are building one at each port. This sort of vessel is 30 feet long, and about 10 over, so constructed with cork, both within and without, as to be incapable of sinking; sharp at each end; she moves either way, merely by the rowers changing their seats. She is navigated by 11 men, and is capable of saving from 20 to 30 persons at each time of going off. By her buoyancy and sharpness of form she can make her way through the most tremendous sea. Thus, under Providence, a firm security is held out to the bold sailor, who has ever been ready to risk his own life, and a greater probability of success is offered in his generous attempt to save the lives of others in the most imminent and horrid of dangers. These vessels are best constructed in the North; the first cost is about 160£ – providing oars, getting home, and building a boat-house, will raise it to about 200 guineas – a provision must be made for repairs; and though the seaman has hitherto shown his generosity by braving death in the horrors of a tempestuous ocean, with no other instigation than that of humanity, yet a generous public will see the necessity of providing some additional rewards; and more particularly in securing a present supply to the families of those who are unhappily lost in the attempt. The great use of this plan will be best illustrated by a single fact: Last January 9 vessels were wrecked off Shields, and the crews, 108 persons, were all saved by the Life Boats, from the most imminent hazard. Scarce a year passes without producing these horrid scenes of distress off the dangerous coast of Lowestoft – the most remarkable perhaps, in the history of man is, that on the 19th of Dec. 1770, when 30 vessels were lost at one time on the sands, in near view of the shore, and the whole crews, consisting of 300 persons, perished to a man, with the most painful circumstances of distress; many of them were seen hanging upon the rigging and the yard arms all the day, and the next morning discovered some few yet in the same situation, without there being the least possibility of assistance. One of these boats might have saved the whole. From the depth and boldness of the shore a boat can put at once to sea without difficulty or delay; and from the establishment of pilots, a hardy set of mariners used to the intricacies and dangers of the coast, are always ready with their services. These circumstances all combine and powerfully call for the establishment of this plan of safety at Lowestoft, as one station.

In full assurance of the liberality of a county, that has ever been forward to relieve distress in whatever form, a vessel has been already ordered to be built without delay, that it may be ready to act the ensuing winter; and may God prevent any circumstance of danger before its arrival. The different Bankes of the county are requested to receive subscriptions, and to transmit the names of subscribers to us.

ROBERT SPARROW
FRANK BOWNESS

4. A contemporary model of the Greathead lifeboat supplied to Bawdsey in 1801 now in the collection of Tyne and Wear Museums, Newcastle upon Tyne. It probably came from Greathead's own workshop and portrays a boat of the same design as the first Lowestoft boat. Reproduced by courtesy of Tyne and Wear Museums Service.

The following extract contains an extended version of the letter just quoted. The early part of the letter is to all intents and purposes the same; the latter part begins with the final paragraph of the earlier letter:

Ipswich Journal 11 October 1800
In full assurance of the liberality of a county, that has ever been forward to relieve distress in whatever form, a vessel has been already ordered to be built without delay, that it may be ready to act the ensuing winter; and may God prevent any circumstance of danger before its arrival. Surely the humanity of the county will not stop here, but will endeavour to provide for the security of their whole line of coast. The next point of danger southwards, offers itself off the Haven at Orford, which place has also the advantage of a body of pilots stationed there; but as those vessels cannot act for a greater extent of coast than 12 miles the distance between Lowestoft and Orford would be too great for any joint assistance; and if that should be thought necessary, an intermediate one might be placed at Southwold or Aldborough – Dunwich would be the best point of distance, if it were sufficiently inhabited by seamen. One of these vessels might be built and established for about 400£, the overplus forming a fund for the purposes before mentioned; but if the public should take up this charity in the manner we trust it will, not less than 1200£ will be necessary, nor is this a sum to alarm us – the benevolence and the opulence of the county is fully equal to it. Some assistance may be expected from the mercantile part of the kingdom; it will therefore be necessary to extend the subscription as far as possible. We earnestly request the ladies, gentlemen, and all individuals, to consider our application in its genuine humanity and to promote it, not solely by their contributions, but in exerting themselves to procure the contribution of others. We also particularly request the aid of the Clergy in their respective and neighbouring parishes, well considering the consolation we shall feel in the hour of the tremendous storm, that we have promoted every human means of alleviating its horrors.

If full success should follow our endeavours, a public meeting will be necessary to arrange the operations. The different Bankers of the county are requested to receive subscriptions, and to transmit the names of the subscribers; a book will be published, in which each parish will be classed in its respective hundred, whether it affords any subscription or not; and every subscriber, whatever may be his sum, will be mentioned, that the names of all those who contribute to the establishment of this, the most humane plan that was ever offered to society, and which may be called the SEA CHARITY, should be handed down to the latest posterity. Subscriptions received at the different Bankers of the county, and at Dorset, Berners, and Co. Bond-street, London.

ROBERT SPARROW
FRANs. BOWNESS

The first recorded lifeboat for Lowestoft, 30ft long and 10ft 6in. in beam, arrived in the town in February 1801. It was double ended and, in theory at least, could be rowed in either direction; the keel was relatively short and the stem and sternpost were long and raking (Plate 5).

Ipswich Journal 7 March 1801
We have the pleasure to inform the benevolent subscribers to the Life Boat, that one of these vessels arrived at Lowestoft, yesterday sen'night; she is 30ft. long, 10ft. wide, sharp at both ends: on Monday 12 men went to sea with her to practise a number of manoeuvres, and in the opinion of the best judges, she is well calculated to answer the purpose intended.

Although similar boats carried out excellent lifesaving work elsewhere on the coast, at Lowestoft the boat proved most unpopular and unsuccessful.[6] The local beachmen who serviced shipping in the Roads[7] and carried out salvage work in the area did not approve of the boat and refused adamantly to use it, saying that it was unfit for use and unsuitable for its purpose.

The boat was quite unlike those used by the beachmen, which were fine sailing craft able to go to wrecks on the offlying sands a considerable way from the shore. The lifeboat was only a pulling boat, and the beachmen considered it totally unsatisfactory for rescue work offshore. While quite prepared to risk their lives in their own 'yolls' when necessary to save others, they saw no point in risking them in a boat that they felt was unsafe.

Although a carriage was obtained so that the boat could be more conveniently moved along the beach and launched at the most suitable place it remained idle in its shed. Robert Sparrow became frustrated with the attitude of the beachmen, and in 1804 he penned a letter calling for a more positive approach to the lifeboat:

Ipswich Journal 13 October 1804
The following letter has been sent to the pilots, sailors, and seafaring men of the town of Lowestoft.
Your late worthy Magistrate, Mr. Bowness, and I had seen with great satisfaction, the exertions made in cases of Shipwreck, and we thought that we could not take a better step than to attempt to raise by subscription, a fund sufficient to establish one or more Life Boats, as a means of rendering those exertions more beneficial to the unfor-

---

6 For a fuller account of this boat see R. Malster, *Saved from the Sea* (Lavenham, 1974), pp.43–46.
7 The anchorages between the shore and offlying sandbanks off Corton, Lowestoft and Pakefield in which vessels took shelter from gales under the lee of the sands.

tunate sufferers and less hazardous to the generous man, who risqued his own life to save that of another. Other gentlemen joined us in the undertaking, and you all know that our zeal was exceeded by the humanity and liberality of the public, and that two Life Boats were established, one at Lowestoft, the other at Bawdsey, in Suffolk. Two circumstances, of contrary natures, attend the situation at Lowestoft, one, that it is very fortunate in the frequency of seafaring men, used to all bodily exertions, and who had already shewn their intrepidity and generosity in assisting distress; the other, that by want of a creek there was a difficulty in getting out to sea, when the winds were much contrary; this difficulty has been in part obviated, Government having supplied a cable and hawser, at the instance of the late Lord Henniker.[8] Whence comes it then that the Life Boat operating with so much success in all other places, is at Lowestoft an object of dislike, that so far from resorting to her when opportunities offer, few occasions are lost to lower her services in the opinion of the public? Motives have been suggested, but they are too disreputable to be believed in any man who has once shewn his fellows that the zeal of saving a shipwrecked sinking sufferer goes beyond his regard for personal safety. Various means have been tried to conquer this unfortunate and unaccountable misconduct. The duty I, as the surviving promoter of the undertaking in this district, owe to the public, calls upon me to make the last effort that remains. You have all along been assured that every attempt should be amply rewarded, and that at the first instance of it, a system of reward should be established. I call your attention to the great success that has attended the Life Boat established at Bawdsey, where in the course of last winter, were saved, at one time, the lives of six or eight persons, who without her assistance, must all inevitably have perished; for though it was in the sight and hearing of numbers upon the shore, no other vessel could possibly have lived at sea. To a generous public this one instance of success of their humanity will be a full reward for the ample sums they have advanced; but it must not cease here, and to endeavour to rouse you from the languid state in which you are, or from the disgraceful prejudice you have adopted; I offer you, from the remainder of the subscription left in my hands, a reward of Ten Guineas for every exertion, fairly and fully made, with the Life Boat, and that if the exertion be attended with the success of saving a human life, the reward shall be extended to Fifteen Guineas. The claim to the reward to be determined by the committee. Our rewards must be proportioned to our means, and should the prejudice against the Life Boat fortunately cease, and she be brought into frequent use, there can be but little doubt but that the same humane spirit which has ever attended the opulent inhabitants of Lowestoft, to reward the brave men who in such circumstances nobly hazard themselves, will continue to be exerted, and it may be hoped other modes may be contrived to prevent the meritorious going without an adequate reward. It rests with you to bring this fortunate system forward; should my present attempt fail, I shall first state the failure to the public, and then look out for a situation where a more worthy race of men will gladly accept the advantages you will have blindly refused, October 1, 1804.

Eventually, convinced that the boat could do no good at Lowestoft, Robert Sparrow moved her to Gorleston, where she could be launched into the harbour and rowed out of the harbour entrance when needed, but she seems to have found no greater favour there. Indeed, when the gunbrig HMS *Snipe* was wrecked in the South Ham at Gorleston in 1807 and more than seventy lives were lost nobody seems to have thought of sending for the lifeboat; she remained in her shed beside the harbour, though she had been designed to deal with just such an incident, a wreck close inshore. It was the wreck of the *Snipe* that prompted Captain George

---

8   Of Thornham Hall, Suffolk.

5. This woodcut shows a Greathead boat being hauled along the beach on a two-wheeled carriage. What was probably a rather similar carriage was made for the Lowestoft lifeboat but, although it must have made launching the boat somewhat easier, it did nothing to popularise the boat with the seafaring population of the town. The woodcut is one of those used by Captain Manby to illustrate his pamphlets publicising his lifesaving proposals, and is in the editor's possession.

Manby (an artillery officer and barrack master at Yarmouth) to develop his life-saving mortar as a means of establishing communication with a wrecked vessel.

There is a probability that the first Lowestoft lifeboat did once go out on service, for a set of accounts[9] published in 1809 includes a payment for 'Going off to a wreck' as well as payments to seamen for certain trials, but it has always been said that she never saved a life. It was only in 1807 that the unsuccessful Greathead boat was replaced by a sailing lifeboat designed on the lines of the local beachmen's yawls (pronounced yolls), and built by Lowestoft boatbuilder Batchelor Barcham under the personal supervision of Lionel Lukin,[10] a London coachbuilder who had earlier experimented with what he called an unimmergible boat and had designed a lifeboat on the lines of a Norwegian traditional boat, a joll (not to be confused with the Lowestoft beachmen's yolls).

Lukin had been staying in Lowestoft that summer and during his time in the resort had discussed with the local beachmen their objections to the Greathead boat. It is said that the pilots' favourite boat, one of the local yawls, was pointed out to him during the discussions, and he was told that if a boat were to be built on the same lines and made unsinkable the local men would gladly use it to save lives. Some of the beachmen wrote to Robert Sparrow indicating that they would willingly use a new lifeboat 'provided we are convinced, by proper experiments, that such a boat is manageable here, and that she will not sink if filled with water, in addition to her proper burthen – this being a degree of safety which we confess our yawls do not possess'.[11]

Sparrow decided, after consulting Captain Gilfred Reed, an Elder Brother of Trinity House who had had charge of the earlier boat, to order the building of a new boat. Lionel Lukin offered to remain at Lowestoft to oversee the building of the boat, which was fitted with a number of sealed tubs or casks lashed inside the boat and under the thwarts to provide buoyancy should the boat be swamped; in later boats these barrels were replaced by air boxes carefully made to fit the internal shape of the hull. Other barrels in the bottom of the boat could be filled with water 'when necessary to encrease her ballast'. The boat also had the protruding cork 'padding' that was to be a feature of future Norfolk and Suffolk type lifeboats.

When the boat was nearing completion Robert Sparrow set down the story of his efforts since 1801 to establish an effective lifeboat service at Lowestoft:

Ipswich Journal 24 October 1807

LIFE BOAT

EVERY Apology I can make is due to the Subscribers to the Life Boat stationed at Lowestoft, for the delay in the detail of the circumstances, and the statement of the subscription, which I have so long owed to them. The pleas I request the public to admit in extenuation, are, first, the loss of my worthy colleague, the late Rev. Mr. Bowness, from whose humanity the attempt had originated; the interruption by the failure of success, and afterwards by some unfortunate events. If the detail had been made at the proper time, I should have had to state, that the boat had been provided with every

---

[9] The editor borrowed a copy of this document from Mr Eugene Ulph and returned it about 1970, but no copy can now be traced.

[10] Author of *Description of an Unimmergible Life Boat* (London, 1820). He originated from Dunmow in Essex and is said to have tested his first model of an unimmergible boat on the Doctor's Pond there.

[11] *Ipswich Journal* 24 October 1807.

thing necessary for her destination, according to the information from Shields, and other places, where these vessels had been used with the most happy success, but that nothing could reconcile her to the pilots and seafaring men of Lowestoft, who were to use her; and that as it appeared the difficulty of getting her out to sea, particularly when the wind was adverse, increased the dislike, I removed her to Gorleston, upon a haven, where she might be launched with great ease in a short time; but that here, also, the same dislike attended her; and that when, in a later terrible storm, the scene of distress was so great, as to induce some humane gentlemen to offer 20 guineas as a reward for the use of the Life Boat, the refusal at once convinced me nothing was ever to be expected from her on this part of the coast. This conduct appeared to me and to others to be the effect of prejudice; and this I should have had to have stated, was unconquerable. I have now reason to suspend this opinion, and to rejoice the statement had not been made; there is a prospect the expectations of the public may yet be realized; some late attempts have been made to establish a mode of relief to vessels in distress with very fair success. Mr. Lukin, a gentleman well known in the cause of humanity, who has made the system of Life Boat his particular study, has written a treatise upon the subject, and received a patent for a Life Boat so long ago as the year 1785, was fortunately at that time resident at Lowestoft, and was present; the many conversations he had with the sailors there, produced the following letter addressed to me:

Lowestoft Septr. 21, 1807

Sir, We whose names are hereunto subscribed, being informed you are desirous that the Life Boat brought hither from Shields should be altered and improved, so as to be more suitable for this coast, or that a new one should be built in such form as we, by our experience, judge most likely to answer the intended purpose, beg leave to assure you, Sir, that our disapprobation of the Shields boat arises from no other cause than her unfitness for this shore, on account of her form, which may be very serviceable on a flat coast, but totally unserviceable on this steep shore, where it cannot possibly be launched through a heavy surf without being filled, and when full of water, becomes so heavy, as to be quite unmanageable. How far any alterations may remove these objections, we are not able positively to determine; but are of opinion, they can be only partially removed, and that a new one may be constructed to answer the purpose much better, both as to safety and management; but whatever boat shall be stationed on this coast as a Life Boat by the subscribers, we hereby declare, that we shall, and we believe all our fraternity will, be ready at all times to use it for the purpose intended, in preference to our own, provided we are convinced, by proper experiments, that such boat is manageable here, and that she will not sink if filled with water, in addition to her proper burthen – this being a degree of safety which we confess our yawls do not possess. We are &c.

Signed by 39 names.

Mr. Lukin also offered to remain purposely at Lowestoft, and superintend either the alteration of the present boat, or the construction of a new one upon his own principles; these he had discussed with the sailors, and I found they approved them; the objections they made to the old Life Boat went to every part of her; the alterations necessary to bring her into favour were calculated at about 100£ and even then the event appeared very doubtful. On the other hand, the pilots pointed out their favourite boat, in which they had made many daring exertions, and declared, if such an one was provided for them, and made secure in buoyancy, and not liable to be overset, they would gladly engage to use her – the expense was calculated at about 200£. A larger sum than this remained in my hands from the former subscription, a statement of which shall be laid before the public. It was not possible for me to call the subscribers together, or by any means to collect their sentiments – the time pressed very much – Mr. Lukin's time was also valuable to him, and I was to make an immediate determination. In this difficulty how to act, the only step I could take was to apply to the Trinity House, who had been liberal in their assistance of money and advice; and the answer of Captain Reed, an Elder Brother of that body, to whom I had addressed my letter, determined me, at my own risque, to hazard the approbation of a liberal public, in ordering a new boat to be built at Lowestoft, similar to the request of the pilots, and according to the principles and under the constant inspection of Mr. Lukin. Should this attempt be attended with the desired success, it will be for the public to determine, how far it will be proper to provide a moderate fund for rewards and for repairs.

Worlingham Hall   ROBERT SPARROW
Oct. 15, 1807

Successful Trials

The new boat was launched for the first time on a decidedly unpleasant November day in 1807 with sixteen people in her. 'Thursday se'nnight[12] was launched the Frances Anne Life Boat, built at Lowestoft under the direction of Mr. Lukin, of Long Acre, London, the original inventor and patentee of unemmergible boats', says a report in the *Ipswich Journal*. 'The weather was very unfavourable, an incessant and heavy rain falling all the day.'[13] The rain might have caused the number of spectators to be rather small, but it did nothing to curtail the boat's trials, the results of which did much to prove that the new boat was likely to prove extremely useful in saving life off the Suffolk coast.

A south-easterly wind was raising a heavy sea on the offshore sands, in spite of which she was taken right across the Corton Sand without her shipping any water. As part of the trial the plugs (apparently more than one was fitted, if the newspaper account is to be believed) were taken up and the boat was allowed to fill with water until she was floating on the air casks, yet she still sailed very satisfactorily in that state, proving very stable. Indeed, when all the people who had gone in her moved to the lee side and some of them climbed on the lee gunwale to test her stability, her trim seemed to be unaffected by their action.

It would have been foolhardy to attempt to beach her in a flooded state, but no pump had been fitted and it was necessary to bail her with buckets after the plugs had been replaced. Her being sailed when full of water is particularly interesting: later lifeboats of what became known as the Norfolk and Suffolk type were normally sailed in that condition so that any water coming in over the gunwale found its way out through the enlarged plugholes, but there is no evidence that this idea had occurred to any of those involved with the boat at that time. It is thought that the principle might have been introduced by Lieutenant Samuel Fielding Harmer, who was second in command of the boat when she was swamped on the Holm Sand in 1825;[14] it is said that at the anniversary of the Suffolk Humane Society in August that year he proposed 'a considerable improvement respecting an increase in buoy-

---

12 19 November.
13 *Ipswich Journal* 28 November 1807.
14 *Colchester Gazette* 29 January 1825; *Ipswich Journal* 29 January 1825.

6. A portrait of Lieutenant Samuel Fielding Harmer, who served in the Frances Ann *with Lieutenant Samuel Thomas Carter and in 1825 suggested certain alterations to the boat to improve her efficiency. The portrait was owned by Miss Teresa Chevallier and was given by her to the Norfolk Nautical Society; it is now in the Time and Tide Museum at Yarmouth.*

ancy in the lifeboat … and the necessary alterations will be immediately resorted to'.[15]

The *Frances Ann* was thus the prototype of a type of lifeboat that found favour with lifeboatmen on the East Anglian coast for many years, the Norfolk and Suffolk type. That type was preferred by the beachmen to the self-righting boats which became standard with the Royal National Lifeboat Institution in the second half of the nineteenth century, and many were turned out by boatbuilders at Lowestoft and Yarmouth in later years. Boats built in the twentieth century had the water ballast confined in a tank rather than swilling loose in the bottom of the boat.

With the building of the new lifeboat the redundant Greathead boat was advertised for sale:

Ipswich Journal 14 November 1807
LIFE BOAT

To be SOLD (her form not suiting the Steepness of the Shore), now lying on the beach at Lowestoft, in Suffolk, built by Greathead, of Sheilds [*sic*], 30 feet long, and 10 feet 6 inches broad, 12 oars, with grummets, and 2 staves, 8 billedge planks for a road, and a carriage with 4 wheels, to move the boat upon. Enquire of Mr. Barcham, boat-builder, Lowestoft.

The *Frances Ann* was built partly from the money remaining from the 1800 subscription and partly from Sparrow's own pocket, but it was the Suffolk Humane Society that would operate the lifeboat service at both Lowestoft and Pakefield for more than sixty years. Exactly when the *Frances Ann* came under the Society's control is unclear as the available evidence is imprecise, but it was probably prior to the 1808 anniversary meeting at which Robert Sparrow, a vice-president, took the chair. The fact that the new boat was named *Frances Ann* after the daughter of Lord Rous of Henham, the president of the Society, is suggestive of an early involvement. As early as 3 March 1806 the committee of the Society had suggested an amalgamation of the lifeboat project with the Society, but discussions with Robert Sparrow did not have the desired result at that time. Possibly when faced with the prospect of heavy expenditure on the new boat, which cost some £350, Sparrow approached the Society for help, but the surviving records do not shed any light on this. What is certain is that in 1809 the *Ipswich Journal* stated that 'Mr. Sparrow has since given up the boat to the Suffolk Humane Society.'

The surviving records of the Society do not include the minute book running from 1806 to 1833 which might have shed light on the beginnings of the lifeboat venture. Happily the gap is filled to some extent by reports in the *Ipswich Journal* and other local newspapers that were written by successive secretaries of the Humane Society. From one of these reports, which appears to have been copied directly from the minutes of the meeting, we learn that at the anniversary meeting in July 1807 members agreed to present the Society's silver medal to Captain George Manby of Yarmouth, the pioneer of the lifesaving mortar, and to Mr Crowfoot 'for his exertions in the case of Serjeant Bubb, which led to the formation of this Institution'.[16]

---

[15] *Ipswich Journal* 3 September 1825.
[16] *Ipswich Journal* 6 August 1808.

The medals were presented to Captain Manby and Mr Crowfoot at the following anniversary meeting.

Ipswich Journal 6 August 1808
On Wednesday last the Anniversary of the Suffolk Humane Society was held at the Crown Inn, Lowestoft, Rt. Sparrow, Esq. in the Chair. The company was highly gratified by the very appropriate addresses of the Chairman to each of the persons to whom medals and rewards were granted. The very valuable invention of Capt. Manby, for securing a communication between the shore and a stranded vessel, was the cause of his receiving a medal with a suitable motto. To Mr. W.H. Crowfoot another was presented, from whose exertions, in restoring the life of serjeant Bubb, the Suffolk Humane Society commenced. Mr. Garrard, of Walberswick, Mr. James Stebbens, and James Buxton, pilots, of Lowestoft, each received a medal, for having at great hazard to themselves, saved the lives of others. Pecuniary rewards were given to Wm. Allerton, Henry Catchpole, and Jas. Stebbens, jun., of Lowestoft; also, to J. Crisp, Jn. Galer, James Cundy, and Robert Sterry, of Southwold. Many interesting discussions, for most effectually preserving the lives of sailors, were during the day introduced, and the company had the pleasure to find the Suffolk Human Society was becoming better known, by the addition of many new subscribers.

Earlier in 1808, on 24 February, the *Frances Ann* was launched on service for the first time when two vessels went aground on the sands to the north of Lowestoft, probably the Corton Sand. The incident was described by the Yarmouth correspondent of one of the local papers:

Norwich Mercury 5 March 1808
Yarmouth
On Thursday se'nnight in the morning two colliers ran aground on the sands off Lowestoft, one of which was got off the same morning and run on shore; the other remained fast till Friday morning when she was got off and brought into this port.
 The wind and tide being against the Lowestoft boats, the Gorleston boats got to the vessel first, but did not attempt to board her till the Lowestoft Life Boat was very near, with the determination of boarding her; which the Gorleston men perceiving and fearful of losing their prize, made the attempt to board her first, in which they succeeded.
 This is the first occasion which has happened to try the Life Boat, which promises to be very useful in rendering assistance at a time when other boats dare not venture out to sea.

On that occasion the lifeboat did not actually save lives, but she had shown her capabilities. In December the following year, however, she did save the carpenter of the *Catherine* when that vessel was wrecked on the Holm Sand, though the rest of the shipwrecked crew were drowned before the boat arrived. The local newspapers contained reports of the lifeboat's service:

Norwich Mercury 23 December 1809
Lowestoft, December 17: Our coast presents a most distressing scene. The beach is strewn with wreck, and between this place and Yarmouth many vessels are on shore. The gale on Thursday evening was tremendous. The extent of the damage it occasioned cannot at present be calculated. One vessel the *Catherine* of Sunderland, coal laden was upset off Lowestoft. From the position of the ship and the violence of the sea, it was impossible for some time to ascertain whether any persons were on the

wreck. At length one man was discovered, and the Life Boat launched as soon as possible, which showed her peculiar excellence in the manner in which she rose on the surf and stemmed the torrent. When the Life Boat reached the wreck the only survivor was almost perished with cold and great was the difficulty which still remained to get him in to the boat. To drag him by a rope through the water was thought too hazardous in his exhausted state; another mode was adopted therefore, but which, in securing the person exposed him to be crushed between a piece of wreckage and the Boat. Some of his ribs were broken and in that state he was conveyed to the shore, where every possible care and attention have been shown to him.

The report of this rescue prompted a letter from an unnamed correspondent who suggested the use of a daughter boat that would provide a means of communication between the lifeboat and a wreck. Such a system has never been adopted in Britain, though some modern motor lifeboats now carry a small inflatable boat that can be used for such a purpose, but in Germany lifesaving cruisers have been built that have a daughter boat in a slipway in the stern.

Ipswich Journal 30 December 1809
Observing some account in the last Ipswich Journal, of a man saved from a shipwreck, off Lowestoft, by the Life Boat of that place, but who had received considerable injury in being taken into the boat, the writer of this article was induced to make some enquiry upon the subject; the result of such enquiry is, that although this Life Boat is perfectly well adapted for the purpose, yet there is still wanted a small boat built on the same principle, to be attached to the larger one, to be either taken in tow or carried within her. The use of such a small boat is obvious and important. Whenever the large one dares not with safety approach the wreck, the smaller one may be veered away, and as the ship wrecked seamen get into it, may with facility be hauled to the great boat, and if on the late occasion there had been such a boat the accident would not have happened; this does not rest on theory, it hath been successfully tried by the Lowestoft yawls and their punts. The public is indebted, for the present Life Boat, to the laudable exertions of Robert Sparrow, Esq. who, when the old Life Boat was found to be useless for this coast, built the present boat from the remains of the old subscription, himself supplying the deficiency; and Mr. Sparrow has since given up the boat to the Suffolk Humane Society. But to keep the present Life Boat and her appendages in a proper state for use; to provide for accidents, and in some occasional and strong cases for rewards, and to build the small boat alluded to, will involve an expence disproportioned to the funds of the society, and it is become necessary, by another appeal to the public, to carry into complete effect, by a second subscription, what, from a train of untoward circumstances, their first has proved unequal to. And it is to be hoped and trusted, that an appeal, which has for its object to shorten the duration of human suffering, and the preservation of human life, will soon be made, and meet with the success it so well deserves.

The survivor from the *Catherine* was badly injured during the rescue, but his recovery was announced in the local papers a few weeks later:

Norwich Mercury 20 January 1810
The conduct of the sailors who saved the carpenter from the wreck of the Catherine a few days ago received the thanks of the Committee. The exertion and humanity of all on board the Life Boat on that occasion could not be too highly praised. We rejoice to add that the carpenter is recovered from the injury occasioned by the wreck and is now enabled to return to his friends.

The new lifeboat had indeed shown how useful she might be, but there remained still a prejudice against such craft among some of the seafaring community. Those who gained their livelihood from attending on ships in trouble naturally preferred to use their own yawls whenever they could, for the gentlemen who ran the lifeboat stipulated that it should not be used for salvage work – and that was the beachmen's business. The italics in the report below are in the original:

Ipswich Journal 17 November 1810
Lowestoft, Nov. 14: – On Saturday morning early a most tremendous gale visited our coast. The eastern side of the town of Lowestoft has suffered very materially by the unroofing of houses, the falling of chimneys, and the breaking in of windows. At Corton 3 vessels are on shore; the crew of one, consisting of 8 persons, was saved by Captain Manby's mode of communicating with the vessel by firing a grapnel. The other two crews were saved. On the sand opposite Lowestoft a vessel struck, and the crew were seen in the yards for a long time; the life boat was hauled down, but the sailors, so great was the danger to be apprehended, were unwilling to make any effort to approach the wreck, although twenty-five guineas were offered to them by some gentlemen who were spectators of this distressing scene, alleging that the danger to be apprehended was very great, and *that they had not been rewarded for former services.* It is, however, to be observed, that the best sailors were engaged at Corton.

The Secretaries

One of the first secretaries of the Society (there were two apparently joint secretaries) was the Revd William Tylney Spurdens, who was at the time chaplain to Lord Rous at Henham. Spurdens was one of the first to make a study of the East Anglian dialect. Though not given the credit he deserved when the book was published in 1830, Spurdens made a considerable contribution to the Revd Robert Forby's *The Vocabulary of East Anglia* (London, 1830), having with a friend begun a collection of what they termed *Icenisms* as early as 1808. Some years after the posthumous publication of Forby's book, Spurdens produced a manuscript for a supplementary volume which was published in 1858.

The other secretary at the outset was the Revd Michael Maurice, who lived at Normanston on the Beccles Road out of Lowestoft. He it was who took the minutes of the inaugural meeting at Kessingland King's Head on 7 January 1806. Both he and Mr Spurdens signed the minutes of the second meeting.

Which of them supplied the *Ipswich Journal* with the report of attempts to save the crew of a Dutch vessel that was run ashore at Kessingland is not known, but from the style it would seem very likely to have been one of the secretaries. In this case it was not the lifeboat that was employed but the lifesaving apparatus that was also in the hands of the Humane Society.

Ipswich Journal 20 January 1810
The following very interesting narrative was presented to the Committee of the Suffolk Humane Society, held at the Rev. J.G. Spurgeon's,[17] Lowestoft, on Monday, Jan. 15, 1810. Its importance to the public does not merely depend upon its authenticity, but also upon the knowledge of the means that have been effectual in preserving life, when

---

[17] The Revd John Grove Spurgeon (1747–1829), rector of Oulton and Clopton, was an early subscriber to the Suffolk Humane Society and a vice-president.

exposed to the greatest danger: – on Saturday, Jan. 13, the hoy *Elizabeth Henrietta*, of Papenburgh, Capt. Vanderwall, from Liverpool to Rotterdam, sprung a leak, and after 15 hours of incessant toil at the pumps, the men were obliged to run the vessel on shore, near the signal house in Kessingland: the distance from Lowestoft is near 4 miles; the wind was at East, and blew very strong; a very heavy surf was upon the shore; it was evident that unless a communication could be secured by throwing a line from the shore to the ship, according to Capt. Manby's judicious plan, the crew must inevitably perish; all the apparatus was at Lowestoft; every possible exertion was applied to facilitate its removal; still, from the distance, the nature of the cliff and the roads, unavoidable delay occurred before the mortar could be fired. In the mean time a buoy was veered from the ship, but not near enough to be reached by a grapnell. The crew consisted of the Captain and 7 men. The Captain betook himself to the shrouds about 3-quarters of the way up the mast; the 7 men secured themselves on the bowsprit. The deck was underwater; the whole ship ready to sink. In those circumstances the mortar was fired; the shot and line reached the bowsprit, and fell in the midst of the 7 men. The line was only one inch and a half in circumference. To this, the 7 men fastened themselves, about 3 yards distant from each other. They then dropt in succession into the sea, and sunk till the line was hauled tight from the shore. Sometimes they were seen, sometimes covered with the sea. In this manner they were dragged about 80 yards through the water, and then all safely landed: 6 out of the 7 lowered themselves into the sea, free from entanglement; the 7th, by accident, threw himself on the wrong side of the rope attached to the bowsprit. In this situation they would have perished, had not the rope fastened to the bowsprit broken, when the line from the shore was hauled tight. The feelings and painful anxieties of the people on the shore, who were aware of the extent of the pending calamity, can better be imagined than described. The afflictive part of the narrative remains to be stated. Capt. Vanderwall was still in the shrouds, and saw all his people safe on shore. The signs he made shewed the anguish of his mind. All was done for his relief that could be done. A second shot was fired, and the rope attached to it was thrown on the yard of the ship where the Captain was standing. He looked earnestly at the rope, but from some cause, made no attempt to reach it. The deck was then broken up, and all communication with every other part of the ship was cut off. Another shot was fired, and the rope passed very near the unhappy sufferer. At this instant all the masts gave way, and the Captain was buried in the midst of the wreck. The greatest praise is due to the pilots and seamen of Lowestoft and Pakefield, for their zeal and exertions upon this occasion; their zeal and exertions are indeed, on all occasions in which distress is witnessed, a most honourable part of their character. It would, however, be highly improper to omit the mention of Robert Rede, John Denny, Charles Barrel, Thompson Swan, James Stebbings, jun., Lowestoft; John Cunningham, of Pakefield; and John Davy, of Kessingland. The activity and perseverance shewn by the above persons, cannot be too highly extolled. The judicious plans and indefatigable attention of Captain Hunter, of the R.N., Mr. James Reeve, and Mr. Elph, entitle them to the highest approbation.

A later secretary was the Revd Bartholomew Ritson, curate of Lowestoft,[18] whose memorial is to be seen in St Margaret's Church at Lowestoft. In January 1815 he contributed a report to the *Ipswich Journal*, reproduced below, of the saving by the Lowestoft lifeboat of three men from the billyboy sloop *Jeanie* of Hull, which had grounded on the Holm Sand.

---

[18]   He was curate of St Margaret's, Lowestoft, for thirty-seven years and also perpetual curate of Hopton. A large silver cup was presented to him by the parishioners and was bequeathed by him to the vicar and churchwardens in his will of 7 March 1835. The 'Ritson Cup' is now in the Norwich Cathedral treasury.

Ipswich Journal 21 January 1815

To the Editor of the Ipswich Journal
Lives saved by means of
the LOWESTOFT Life Boat

Through the medium of your very useful and extensively circulated paper, I beg to give publicity to the following recent instance of intrepidity and courage exhibited by some young men belonging to this town, who were providentally the means of saving three fellow creatures from a watery grave. On Friday the 13th inst. at day-break some of the Lowestoft boatmen being on the look out, perceived a wreck lying among the breakers on the Corton sand, otherwise called the home sand, the wreck bearing ESE from the Lowestoft Upper Light-house, and distant from shore about two miles. Three of the pilot yawls were soon manned and put off to visit the wreck to ascertain whether there were any persons on board, and if so, to render whatever necessary assistance might be in their power. Upon approaching the wreck, the people in the yawls discovered three people on it, but at the same time found, to their great mortification that by reason of a tremendous sea upon the sand, and the high surf and broken water surrounding, and frequently breaking over the wreck, it would be impossible for any yawl to get nearer without manifest hazard of being dashed to pieces. A signal being thrown out to shore by one of the yawls that there were persons on the wreck, the life boat was got out with the utmost expedition, and as soon as upon the beach, was instantly manned by the following persons, viz. Henry May, David Burwood, James Cullingham jun. and Henry Beverley Disney (pilots), Cornelius Ferrett, William Ayers, Samuel Spurden, John Spurden, Robert Watson, James Websdale jun., Samuel Butcher jun., Bartholomew Allerton, James Farrer jun., Peter Smith, George Burwood, Matthew Colman, Edward Ellis jun. and James Stebbens. The alacrity with which these brave fellows leaped on board the life boat is scarcely to be described; after encountering much difficulty and danger in passing the breakers they came near the vessel in sight of hundreds of spectators, who, from the heights were beholding with astonishment their admirable nautical skill and dauntless courage; at the same time trembling between hope and fear for their safety, and lifting up a silent prayer for the successful termination of their perilous undertaking. Heaven in its mercy smiled propitious on their endeavours, and rewarded the exertions of these brave men with success, and they had the heartfelt joy of bringing the three shipwrecked mariners to shore without any accident. The names of the men saved from the wreck are James Naylor, master, Abraham Chitson, and John White, seamen. The sloop proves to be the *Jeanie* of Hull, laden with potatoes, and bound to London. She sailed from Hull on Thursday morning, and about twelve at night, when off Hasborough Gat, sprung a leak, which gained so fast upon the crew that they were obliged to run on the sand to prevent her foundering. Those gentlemen resident in Lowestoft, who are members of the Suffolk Humane Society, to whom the life-boat belongs, with a promptitude which does honour to their feelings assembled, immediately after the transaction, to form a committee of reward; and having first resolved that the above was a case which came within the views of the society to reward, they unanimously voted that the sum of five guineas be paid from their fund to the Life-boat's company, as a small token of the high estimation in which their humane and meritorious conduct was held. The gentlemen constituting the committee were the Rev. John Grove Spurgeon, Vice President, the Rev. Richard Lockwood, Treasurer, the Rev. Bence Bence, Treasurer at Beccles, Robert Reeve Esq., Robert Reeve jun. Esq., Mr. John Elph and Mr. Alexander Payne. In the case above stated, a most favourable opportunity has been afforded to the Suffolk Humane Society of proving to the public the comparative safety and consequent utility of the Lowestoft Life Boat, and with it has been happily effected, as far as human means can avail, that which could not have been effected by any other boat from this beach; and what is most desirable, all doubts respecting its eligibility as a sea boat, have now been cleared up, and all

xl

former prejudices to its disadvantage, which have long been known to have existed in the minds of our seamen, removed. The Suffolk Humane Society have further to hope that the present instance will be the means of securing two other very important ends; – First an increase in the number of its members and benefactors, and secondly, the firm and undeviating support of those whose names are already enrolled as subscribing members; for as the society has dependence upon no fund, but individual liberality, it must be obvious that it is only by subscriptions regularly remitted when called for, that the annual expences of the Life-boat can be defrayed, and that noble and praise-worthy object which first gave birth to the Institution perpetuated. "The reward of the meritorious". Long as I have retained your readers, I should deem myself wanting in an act of common justice, to a most deserving member of the community, were I to omit mentioning that the Lowestoft Life-boat was planned by Mr. L. Lukin, an eminent coach-maker in Long Acre, London, and built at Lowestoft under his imme-diate superintendence.

I am Sir,
Your very obedient servant   BARTH. RITSON
Curate of Lowestoft, and Secretary to the

Lowestoft, Jan. 19, 1815   Suffolk Humane Society.

Five years later Mr Ritson was the author of a lively report, reproduced below, of a particularly meritorious rescue when the *Frances Ann* saved twelve men from two sinking vessels off Lowestoft who 'were safely landed on the Beach at Lowestoft, without the smallest accident whatever, amidst the congratulatory cheers and greet-ings of the anxious multitude who had been witnesses of the distress':

Suffolk Chronicle 28 October 1820
SHIPWRECKS. – The gale of Sunday morning proved dreadfully and calamitously destructive on the eastern coast. The Rev. Mr. Ritson's letter from Lowestoft, in a subsequent column, announces the loss of 2 vessels, names unknown, with all their crews; and of a third, the *Sarah and Caroline*, of Woodbridge, Jos. Spalding, master, the crew of which were saved only by the astonishing perseverance and exertion of the Lowestoft Life Boat. It was not until the third attempt, when, nearly exhausted, the crew of the Life Boat, assisted by several tradesmen of the town, succeeded in reaching the wreck, at the providential moment for rescuing the wretched crew, who, exhausted by fatigue, and benumbed with cold, were almost incapable of supporting themselves on the rigging. They were in so feeble a state that, on their landing, they were scarcely able to walk to the inn, where every thing was most promptly and humanely supplied for their restoration and comfort. The Life Boat, on her return from the wreck of the *Sarah and Caroline*, rescued the master and crew (8 individuals) of the *George*, of London, which soon afterwards sunk. It is gratifying to know, that Lieutenant Carter, R.N., who had the charge of the Life Boat, with the crew under his command, has received the public thanks of the Suffolk Humane Society.

To the Editor of the Suffolk Chronicle

MELANCHOLY SHIPWRECKS, SUFFOLK HUMANE SOCIETY, &c.

SIR – I shall esteem myself greatly indebted to your kindness by the insertion of the following statement, the correctness of which may be relied on, in your very valuable Paper.

I am Your's,
BARTHOLOMEW RITSON, clk.
Secretary to the Suffolk Humane Society.
Lowestoft, October 24th, 1820.

xli

On Sunday morning last (Oct. 22) a heavy gale of wind from S.S.W. was experienced at Lowestoft, which, towards noon, had increased almost to a hurricane; the whole sea was one continued foam, and the most tremendous surf broke upon the shore. About 12 o'clock, the inhabitants of the town had the pain of witnessing the distress of a vessel which, in attempting to gain the inner roads through the Stamford Channel, struck upon a sand called the Beacon Ridge, and, in about seven minutes, went to pieces and all hands on board perished. A second vessel soon after followed, and in making the same attempt, met with the same melancholy fate, and all the crew were lost! A third vessel, a sloop called the *Sarah and Caroline*, of Woodbridge, laden with coals, struck upon a sand called the Newcome, and remained thereon with her mast standing; but, soon filling with water, the crew, consisting of five persons, took refuge in the shrouds. Here their situation was most perilous; for as it was only half ebb tide, with the wind tremendously strong, no assistance from shore could be afforded them until the following flood, supposing the vessel should hold together so long. – In the mean time, every necessary preparation was made to render assistance as soon as such an attempt should be in any measure practicable. The Lowestoft life-boat, belonging to the Suffolk Humane Society, was got out and manned under the direction of Lieutenant T.S. Carter, R.N., and when launched was towed a considerable way to the southward, to bring her on a bearing with the vessel in distress. Still, however, when the tow was let go the boat fell to leeward of the Wreck; and it was not until the tide began to flow, that she made any way towards attaining her object. After the most persevering and strenuous exertions, however, she succeeded in gaining the wreck, and providentially took the poor fellows from the shrouds, just as one of them was about to drop from his hold through fatigue and cold.

In approaching the sloop, the life-boat passed and was hailed by a brig, coal laden, which, on her return, she boarded, and found her in a sinking state. She proved to be the *George*, of London, John Dixon, master, with seven hands on board. These were also taken in the life-boat; and about six o'clock in the evening the whole twelve persons were safely landed on the Beach at Lowestoft, without the smallest accident whatever, amidst the congratulatory cheers and greetings of the anxious multitude who had been witnesses of the distress. The sloop's mast fell about an hour after the men left it, and the brig sunk soon after.

Too much praise cannot be given to Lieutenant Carter and the pilots, and men on board the life-boat, for their cool, steady and intrepid conduct on this very trying emergency; to whom individually the Suffolk Humane Society have returned their thanks.

About a year ago, those who had the care and conduct of the life-boat were most unjustly and petulantly assailed by AN ANONYMOUS WRITER. The present instance, however, it is to be hoped, will convince the public that all reports to the disadvantage either of the boat or its conductors are totally void of foundation. On this occasion, moreover, has been forcibly demonstrated the utility of life-boats in general, when properly constructed, and of that at Lowestoft in particular; that, as far as human means are available, she is fully adequate to the humane purpose for which she was intended; and that in her, as in the instance now related, may be done what in any other boat or yawl it would be the extreme of rashness to attempt.

Names of the persons who composed the crew of the *Frances Ann* life-boat: Lieutenant T.S. Carter, R.N.; H.B. Disney,[19] pilot; James Stebbens, pilot; James Titlow, pilot; Thomas Aldiss, Henry Smith, Thomas Butcher, William Hook, William Gurney,

---

[19] Henry Beverley Disney was a Trinity House pilot who earned a considerable reputation as a seaman and as a lifesaver, receiving a number of awards for his personal exploits. In 1845 he was asked to take charge of the Lowestoft lifeboat. He was one of those who in 1826 gave evidence to a Parliamentary committee on the Norwich and Lowestoft Navigation Bill.

James Taylor, John Browne, William Francis, Robert Chaston, Edmund Boyce, Thomas Humsley, Nathaniel Killwick, James Robinson, and William Butcher.

Mr Ritson had been able to report a successful rescue, but there were occasions when the lifeboatmen found themselves not merely thwarted in their efforts to save the lives of shipwrecked seamen but at great risk of losing their own lives. Three men were washed out of the lifeboat when it was called out on 18 January 1825, being hauled back on board by other crew members, but in spite of their efforts some of the wrecked men they were trying to save had to be abandoned when the lifeboat was swamped:

Norfolk Chronicle 29 Jan 1825

WRECKS OFF LOWESTOFT

In the dreadful gale of Tuesday the 18th instant, about one o'clock p.m. a brig, the *Ann*, of Shields, with a signal of distress, was observed ashore on the Stanford Sand, off Lowestoft. The Lowestoft Life-boat, commanded by Lieutenants Carter and Harmer, R.N., with a crew consisting of Wm. Folkard and David Burwood, pilots, Geo. Yallop, steersman, and 14 others, was immediately launched, and after being towed to Kirtley[20] Gap, made sail for the vessel; but a strong ebb tide prevented their reaching it. About the same time, another brig, the *Harriet and John*, of Sunderland, the crew of which escaped in their own boat, struck on the Newcome; and a sloop, the *Dorset*, of Ramsgate, on the south-east part of the Holme Sand. The latter being considered in the greatest danger, the Life-boat endeavoured to reach her; not succeeding, she (the Life-boat) anchored. At half-past three, the tide having eased, she got her anchor, and after many attempts, a communication was got with the sloop, then on her beam ends, by means of grapnels thrown to her by Lieut. Harmer. The Life-boat was then hauled under the mast-head of the sloop, on which were the crew, consisting of seven persons; after some time, the master was saved by dropping from the mast, and being assisted into the boat; shortly after another man was saved in a similar manner. At this period the crew of the Life-boat were in a most perilous situation, she being completely under water. One of the many tremendous seas which broke over the wreck upon the boat, washed out of her Lieut. Carter, D. Burwood, pilot, and another, but they were happily saved. It now appeared to every one in the boat, that they could be of no further use, the boat being washed from the wreck, and in such a state, that it was doubtful whether she could reach the shore. They were, therefore, reluctantly obliged to cut the cable, and leave the remainder of the crew of the sloop to their fate. After the Life-boat crew had secured themselves, with great difficulty the foresail was hoisted a little way up, when the boat was happily found to answer her helm, and veered before the sea which was breaking considerably over the heads of those abaft, making its way forward; the one or other end of the boat being alternately under water, so much so, that the crew could not at times see each other. After passing the Newcome Sand, the boat reached the shore full of water, about seven o'clock, when the crew were assisted out of her, so much exhausted as to render it requisite that they should be led or conveyed to their homes.

Very shortly after the Life-boat was launched, the yawl called the *Trinity*, of Lowestoft, with a crew of 16 persons, Wm. Furrell, steersman, was launched and towed along the beach, till unfortunately her outrigger was broken by another boat coming in contact: she was consequently obliged to put back to the main; where, being

---

[20] The modern spelling of the name is Kirkley, but in this report it is referred to both as Kirkley and Kirtley, which were once alternative spellings.

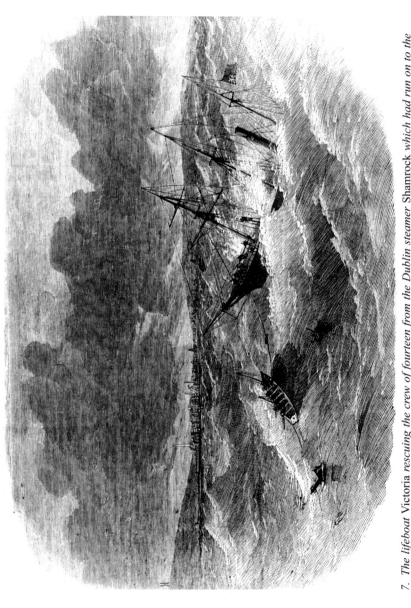

7. The lifeboat Victoria rescuing the crew of fourteen from the Dublin steamer Shamrock which had run on to the south-east part of the Holm Sand on 1 November 1859, a rescue that won Robert Hook and six other lifeboatmen the RNLI Silver Medal. This steel engraving was used in the Illustrated London News of 26 November 1859.

fitted with another outrigger, she was immediately re-launched and towed towards Kirkley by men and horses. About half-past two, observing the life-boat endeavouring to reach the sloop, she cast off and made sail for the brig *Ann*, which she reached, after striking on the Newcombe Sand, and getting over with great peril and difficulty, and was brought to, about a cable's length to leeward of the brig; from which a buoy was immediately veered to her, the sea running at the time so high, that the crew of the yawl dared not venture to lay her alongside the brig; they therefore got their anchor, and having set their foresail, let the anchor go again, and veering out their cable, hauled by the rope from the brig, under her bows, and succeeded in taking the crew, consisting of the master, seven men, and two boys, from the bowsprit of the vessel.

Too much praise cannot be given to Lieuts. Carter and Harmer, the pilots, and crew of the life-boat, and the crew of the yawl, for their perilous and persevering exertions on this occasion.

The sloop soon afterwards went to pieces, and the brig filled and sank. The medals of the Suffolk Humane Society have been awarded to Wm. Folkard and David Burwood, pilots, and George Yallop, steersman of the life-boat, and to Wm. Furrell, the steersman of the yawl, as a small token of the value the Society entertain for their laudable exertions.

At the anniversary meeting of the Humane Society on 25 August that year 'many useful experiments were put in practice under the particular direction of Lieut. Harmer, by whom a considerable improvement respecting an increase of buoyancy in the Life Boat was proposed, and its utility clearly demonstrated'.[21] Unfortunately the report does not tell us exactly what the improvements were that were 'immediately resorted to'.

Sometimes a lack of success led to uninformed criticism of those responsible for operating the *Frances Ann*. In 1825 Edmund Norton, who had taken over from Mr Ritson as secretary of the Suffolk Humane Society, found it necessary to refute allegations of mismanagement when the lifeboat failed to arrive in time to save the crew of the brig *Rochester* which had stranded on the Corton Sand.

Lowestoft, Nov. 3
To the Editor of the Norfolk Chronicle
SIR – A paragraph having found its way into several newspapers reflecting on the conduct of the persons in charge of the Lowestoft Life Boat, and insinuating that in consequence of "*some mismanagement*" the lives of the persons on board the brig Rochester, of North Shields, which was wrecked on Corton Sands on the 20th ult., were lost, the Committee of the Suffolk Humane Society to which this boat is attached, request you will insert the following statement of facts. I am, Sir,

Lowestoft,	Your obedient humble servant,
2nd. Nov. 1825	EDMUND NORTON,
	Secretary to the Suffolk Humane Society.

A vessel being observed on the Corton Sand on the morning of the 20th. Oct. last, the wind blowing a gale from the N. and N.N.E., information was given to the persons in charge of the Life Boat about seven o'clock in the morning, and she was with the greatest dispatch got down to the beach and manned. It being supposed that she could with the assistance of a strong ebb tide beat down to the wreck, she was launched at half past seven, but, after several attempts, owing to the wind shifting to the N.N.W.

---

[21] *Norfolk Chronicle* 3 September 1825.

it was found impossible (although she had beaten some way to windward) to reach the vessel, she therefore made for the shore, and having taken on board a tow rope was towed by eight horses down to Corton rails where she cast off, and before eleven o'clock reached to windward of the wreck, but unfortunately too late to render assistance, as the masts had fallen overboard, and the crew had found a watery grave.

Altho' the exertions of those on board the Life Boat were not crowned with success, they have the satisfaction of knowing that they used every effort in their power and acted to the best of their judgment.

Samuel Carter

Lieutenant Samuel Thomas Carter, RN, who is mentioned in Mr Ritson's report, was a member of a well-known Sudbury family who settled in Lowestoft and became a very successful 'Captain of the Life Boat' or in present terminology coxswain. He was also a member of the Humane Society committee for many years. He had joined the Royal Navy as a first-class volunteer in 1800, attaining the rating of midshipman later the same year. His career afloat was not undistinguished and certainly not entirely uneventful, but it seems to have come to a premature end in 1813 when he was placed on half pay, having been confirmed in the rank of lieutenant in 1808. The year after his naval career had ended Carter married his cousin, the daughter of a rector of Margate, and after their marriage the couple settled in Lowestoft. It was not long before Lieutenant Carter became involved with the lifeboat. By the time he retired at the age of fifty-seven in 1842 he had been instrumental in saving 124 lives. In 1831 Carter and Samuel Fielding Harmer, who in present-day terms would be described as second coxswain and had just been appointed Chief Officer of Coastguard in his native Yarmouth, were presented by the Suffolk Humane Society with silver snuff boxes in recognition of their outstanding work with the lifeboat.[22]

Lieutenant Carter received a severe injury to his right arm in 1836 during an operation in which the *Frances Ann* had to be dragged three-quarters of a mile along the beach. He had recovered by November 1837, when he took the *Frances Ann* to the wreck of the Newcastle trader *Bywell*, rescuing the crew of ten and four passengers, two of them women.[23] The subscribers to Lloyd's voted him their gold medal 'as a mark of the sense entertained by them of his meritorious and humane conduct' in saving the *Bywell*'s crew and passengers.[24]

Norwich Mercury 4 Nov 1837
Lowestoft, Nov. 2
On the 1st instant, at 7 pm, a brig was reported to Lieutenant Samuel Thomas Carter, in charge of the Lowestoft Life Boat, to have grounded on the SW end of the Newcome Sand, shewing night signals for assistance, two of the Beach Yawls attempted to launch, but failed. Lieutenant C. with the valuable assistance of Edmd. Norton, Esq. and many of the towns people, succeeded in getting the Life Boat off, when Lieutenant C. accompanied by Mr. H.B. Disney, trinity pilot, proceeded to the above vessel, and after three attempts, succeeded in taking out the Master, Wm. Banfield, and nine men

---

[22] For a full account of Lieutenant Carter's career, see R. Malster, 'Suffolk Lifeboats – The First Quarter-Century', *The Mariner's Mirror* vol. 55 (1969), pp.263–80.
[23] *Norwich Mercury* 4 November 1837; *Suffolk Chronicle* 4 November 1837.
[24] *Norfolk Chronicle* 23 December 1837.

xlvi

of the *Bywell* of Newcastle, from London, with a general cargo, together with four passengers (viz.) Mr. and Miss Richardson, Miss M.A. Basket, and Mr. George H. Wilson. Lieutenant Carter gives great credit to Mr. Disney and the boats crew, for their very efficient services. Shortly after the Life Boat left her, the brig fell over on her beam ends, and at day-light this morning was a complete wreck, her cargo floating in all directions.

It is interesting to see that Lieutenant Carter praised the crew and Henry Beverley Disney, who had taken over as second coxswain, to use the modern term, when Lieutenant Samuel Fielding Harmer left for Yarmouth. Like all good leaders he was quick to praise his men. He and the *Frances Ann* were in action again on 10 January 1838, putting off to one of the Leith smacks that ran a passenger service from Leith in the Firth of Forth to London.

Norwich Mercury 27 January 1838
Lowestoft, Jan. 25
On the 10th inst. a Leith smack was observed at anchor, close under the Barnard Sand, having a signal of distress flying. At ten o'clock the life boat, under the command of Lieutenant Carter, was launched, and proceeding to the vessel, which proved to be the *Sir Walter Scott*, of and from Leith, bound to London. She had lost her rudder and was making a great deal of water; the wind at the time was blowing hard at ENE accompanied with snow. On finding the vessel in no imminent danger, and that the captain and crew were not disposed to leave her, Lieutenant C. made a signal to the *Seaflower* yawl to remain by her, and taking out a female and a boy they landed at Benacre, where the life-boat was left, at the request of the Captain, in the event of her assistance being required.

The Pakefield Lifeboat

Lieutenant Carter was a member of a committee appointed in 1839 to oversee the building of a lifeboat for Pakefield which also came under the control of the Humane Society. The boat was built at Yarmouth by William Teasdell for £194 1s. 5d., the increase on the tender price of £180 apparently being on account of additional items ordered after the boat was built, including an additional two pairs of oars.

The men who formed the crews of the two lifeboats were pilots and beachmen, the latter being members of the beach companies who made their living by attending on shipping in the Roads and by performing salvage services. Though the beachmen pointed out that salvage was a perfectly legitimate service for which they were entitled to receive their reward, the clergymen and gentlemen who contributed their subscriptions to the Suffolk Humane Society and sat on its committee were determined that the lifeboats should be used only for lifesaving purposes and not for salvage. If the beachmen wished to help get a ship off the sands or to bring a leaking vessel into harbour and to seek a return for so doing, then they must use their yawls for the purpose.

With the Lowestoft lifeboat under the firm control of Lieutenant Carter there was little chance of the *Frances Ann* being employed in salvage work, though this did not avoid friction between the operators of the lifeboat and the beachmen on occasion. In 1821, for instance, there was a most unfortunate incident when one of a party of beachmen who had boarded the brig *Westmoreland* cut the line by which the *Frances Ann* was made fast to the vessel. This case led to the beachmen being charged at the Admiralty Court in London with conspiring to prevent assistance being given to a wreck, by which four lives were lost.[25]

This has led to the oft-repeated statement that the beachmen resented the introduction of lifeboats, a statement that can only be regarded as a half-truth in view of the fact that it was the beachmen who manned the lifeboats. There is no doubt, however, that the Lowestoft beachmen were a stubborn and at times unruly race of men: when the harbour company's tugs began to impinge on their means of livelihood in the 1850s they retaliated by bombarding the tugmen with beach cobbles, and in 1882 their grievances led them to refuse to man the lifeboat, resulting in the 'Black Saturday' incident of 28 October that year.[26]

One can sympathise with them to a considerable extent, if not with their actions. As one of the beachmen said in a letter to the local newspaper in 1882, 'We are allowed to go into danger for the good of others, and it may be to sacrifice our own lives ... Then we are brave fellows; but if we go out to vessels to prevent them getting into dangerous positions and obtain some remuneration for our services, then we are branded with the title of "money grubbers" or "longshore pirates".'

In the case of the Pakefield lifeboat it proved impossible to exert the same strict control, and more than once the committee remonstrated with those in charge of the boat, who were 'desired and directed not to use the boat for salvage in future on any pretence whatever'.[27] Injury was added to insult on one occasion when the committee was told that the boat had been damaged while being used in a salvage operation.

New Lifeboat

By 1849 the *Frances Ann* was worn out, and during one of the tests made at the Lowestoft Roads Regatta that summer she capsized through some of the air casks lashed inside the starboard gunwale having become leaky. It seems likely that the whole of her structure had undergone such strain in services that she was no longer suitable for lifesaving work, and the committee recommended that she be replaced by a new boat.[28] At the annual meeting on 20 September 1849 it was agreed that the building of a replacement lifeboat should go ahead under the supervision of certain members of the committee appointed at that meeting.

In order to save money it was agreed that the spars and sails of the old boat should be transferred to the new one, which was probably an unsatisfactory compromise as the new boat was somewhat larger than the *Frances Ann*. Little is recorded in the minutes concerning the building of the new boat, but at the 1850 annual meeting it was agreed to name her the *Victoria*.

In 1855 it was decided that because of difficulties being experienced with launching the *Victoria* from the beach the boat should be moved into the outer harbour and the directors of the Eastern Counties Railway Company, which was the owner of the harbour, were approached for permission to keep the boat there.

---

25 *Suffolk Chronicle* 15 December 1821; *Suffolk Chronicle* 22 December 1821; *Colchester Gazette* 19 January 1822; *Colchester Gazette* 26 January 1822; *Ipswich Journal* 26 January 1822; Malster, *Saved from the Sea*, pp.51–53.
26 Malster, *Saved from the Sea*, pp.103–16.
27 Committee minutes, 26 January 1852, 4 April 1853.
28 Committee meeting, 6 September 1849.

The directors replied giving permission for the boat to be stationed in the harbour 'provided it is no expense to the company'. First of all there was a quibble over the proposed site of a boathouse, and then the Suffolk Humane Society argued that the railway company should build the boathouse and rent it to the Society since it would be on railway company land. This the railway company steadfastly refused to do,[29] and eventually it was agreed that the lifeboat should be kept afloat in the outer harbour during the ensuing winter.

In the opinion of the committee of the Suffolk Humane Society the railway company was being parsimonious in refusing to undertake the building of a boathouse, but it has to be borne in mind that at the time the Eastern Counties company was in a state of turmoil. After the company had assumed responsibility for the whole Eastern Counties system in November 1854 the engineer, Peter Bruff, reported in February 1855, just as letters were passing to and fro between the Humane Society and the railway company, that at least £150,000 would be required to bring the bridges and permanent way on the Norwich–Cambridge–London line up to standard.[30]

It is perhaps not surprising that the Humane Society should endeavour to persuade the Eastern Counties Railway to pay for the building of a boathouse, for the Society was at that time in the throes of a funding crisis. Subscriptions and donations were proving quite inadequate to pay the Society's outgoings, in spite of the efforts of the officers to attract funds.

At a general meeting of members at the Crown Hotel, Lowestoft, on 18 February 1855 a letter from Captain Ward, secretary of the Royal National Lifeboat Institution, was read stating that a proposal to link the Humane Society and the RNLI would 'be favourably entertained by the committee'. And at a subsequent general meeting on 5 March it was formally resolved that the Humane Society should unite with the RNLI, which from then on was responsible for paying rewards to the crew while leaving the day-to-day running of the station in the hands of the local society.

The Society continued to run the lifeboats at Lowestoft and Pakefield most effectively for some years as well as rewarding individuals for the saving of life in the north-eastern part of the county. It was only in 1873 that the Suffolk Humane Society passed over control of the Lowestoft and Pakefield lifeboats to the national institution.

The Society thus played an important role in the establishment of a lifeboat service on the northern part of the Suffolk coast and for sixty-five years took a leading part in the development of lifesaving. Its history is therefore of national rather than merely local interest.

---

[29] Minutes, 4 January 1855, 12 February 1855, 10 September 1855.
[30] Hugh Moffat, *East Anglia's First Railways* (Lavenham, 1987), p.205.

THE MINUTE BOOK OF
THE SUFFOLK HUMANE SOCIETY

1806

SUFFOLK HUMANE SOCIETY
VOL. 2
MINUTE BOOK
1806

The first 20 pages of this Vol.
contain a Copy of the History
of the formation of the Society
in 1806 from Vol. 1

Then Minutes of Meetings
from July 31 1833
to May 23 1859

No mention of
Miss Crowe

[*There are two blank sheets of paper watermarked BW 1803. The following five feint ruled sheets bear a watermark of Britannia within an upright oval, the oval crowned. The next six sheets are watermarked C Brenchley 1804, and thereafter the sheets are a mixture of those two.*]

[1]

1806

Minutes of the

Suffolk Humane Society

Copy of a letter from Mr. William Henchman Crowfoot,[1] Surgeon at Beccles, to Dr Hawes, Treasurer of the Royal Humane Society, of London.

Sir   Beccles Decr. 7th. 1805
My apology for troubling you with the following narration, must rest upon the tendency it appears to me to have, to promote the ends for which the Society was instituted, over which you preside with so much ability.
A transport with a company of the 28th Regiment of Infantry[2] on board, was wrecked about 11 o'clock on the night of the 18th Inst.[3] on the shore of Kessingland, a village about 8 miles distant from this place. The soldiers and crew, after

---

[1] See above, p. xxii.
[2] Later, under Cardwell's reforms, to become the Gloucestershire Regiment, which amalgamated with the Royal Hampshire Regiment in 1970. See above, p. xxi.
[3] Error for 16th.

experiencing great hardships, were with difficulty landed about nine or ten o'clock the next morning, and the former were marched immediately to Lowestoft. I was accidentally at Kessingland on the morning of the 17th inst. visiting a patient: on my return from whose house I met a cart with some luggage saved from the wreck; & another following it, containing the body of a soldier supposed to be dead.

[2] On my inquiring into the circumstances from the serjeant who had the care of the luggage, he gave me the following account, which I have since had confirmed by the captain of the vessel.

The man in the cart, who was also a serjeant in the regiment, had, about 11 o'clock on the night before, from the effects of cold, & as they supposed, of intoxication, sunk apparently lifeless on the deck. From the hurry & confusion incident to this dreadful situation, & their supposing him dead, he had been suffered to remain there, with the sea continually washing over him, till the morning. When some boats, at infinite hazard, went from the shore to bring off the sufferers, this man had been let down from the vessel by a rope into one of them and brought to land, where he remained two hours lying on the beach. He was left there by his commander; when the humanity of four peasants induced them to place him in a cart provided for the removal of the officer's luggage; but the serjeant who had the care of it, thinking a dead body an unnecessary incumbrance, had him removed. The peasants however procured another cart, on which they were conveying him to the village when I met them.

Upon examination I found no symptom of remaining life, except a very slight warmth [3] about the praecordia.[4] This however encouraged me to attempt the poor fellow's recovery; and I desired them to drive to the inn. It was now about noon; and the man had been supposed dead thirteen hours. I had him stripped, wiped dry, and put in warm blankets: and after three or four hours unremitted perseverance in the means which I judged most likely to restore suspended respiration, I had the inexpressible satisfaction of finding I had succeeded in rescuing a fellow-creature from premature death and in preserving to his country a robust young soldier. During all this time I received every assistance from the four peasants, who had from the first so meritoriously exerted themselves in the cause of humanity: and although they could not be persuaded my endeavours would prove successful, their confidence in me, whom they all knew, induced them to persevere. I will beg leave to suggest that they are poor men, to whom any small reward would be acceptable: and on a coast where accidents are so frequent, a trifling compensation for their trouble might be attended with a good effect.

I am, Sir, &c.
W.H. Crowfoot

[4]   Dr. Hawes, to Mr. Crowfoot
                              Spital Square, December 27th 1805
                              "Breve et inreparabile tempus
                              "omnibus est vitae: sed famam extendere factis;
                              "hoc virtutis opus."[5]

---

4   The area of the heart.
5   'Life is a short and unequal time for all, but able to be extended by deeds; this act of worth.' *Aeneid*, X, lines 467–69.

4

Sir

The communication of your very extraordinary success in the grand and sublime cause of resuscitation, attested by Mr Maurice, and which is also corroborated by my much-respected friend Captain Reed, prompts me to say that <u>Fame extends to deeds</u>: therefore your deserved fame and meritorious exertions in the cause of Life will be recorded in the annual report of the Royal Humane Society as the labour of virtue.

After this introduction I am sure you will be disposed to favour me with a more particular account of the body on your first inspection, the various methods employed, the earliest signs of returning life, and what time elapsed before his perfect return of health, as well as any appropriate reflections on apparent death or resuscitation that may strike you at the time of writing your answer as such will add to your professional reputation.

I am anxious – very anxious that you should be presented with the Honorary Medallion in the most respectful manner, therefore I acquaint you that on the day of the anniversary festival, [5] they are presented by the noble chairman to the successful resuscitative practitioners: but as I apprehend the great distance &c. will prevent your being present at the London Tavern (the latter end of April) it will be doing well to mention some London friend to attend and produce your letter on so important an occasion: and that friend may address you, "<u>Palmam qui meruit ferat</u>".[6] May health and happiness be your constant companion for many years, is the ardent wish of

Wm. Hawes.

Mr. Crowfoot's 2d. letter to Dr. Hawes

Sir   Beccles January 2nd. 1806

I beg you to accept my best thanks for your very polite letter. The desire you are pleased to express for my being presented with the Society's Honorary Medallion cannot but be highly gratifying to me. It will give me great pleasure to inform you of the particulars of Serjeant Bubb's case, since the promulgation of them may tend to encourage others not too hastily to give up an apparently dead person for lost even under the most discouraging circumstances.

If I cannot be so accurate as I could wish as to the precise time which elapsed [6] before any signs of returning life could be perceived you must have the goodness to recollect I was in an obscure village, without any other assistance than what was afforded by the four awkward but humane peasants whom I mentioned in my former letter. Under such circumstances I was compelled to take the most active part, and consequently had not time to be very exact in my observations. My means of restoration were also necessarily extremely limited: but they were employed with zeal and perseverance. They were such too as could be equally well made use of by any person, not of the profession: and perhaps this constitutes the chief value of the case.

The serjeant had fallen down in a state of insensibility about eleven o'clock in the night of the 16th. December, and it was not till noon the next day, that I accidentally saw him. On my removing him from the cart, on which the villagers were bringing

---

[6]   'Let him who deserves it bear the palm.'

5

him from the shore, I could feel no pulse, nor perceive any respiration. There was a slight warmth above the epigastrium,[7] but every other part of the body was cold as marble. No external injury was to be discovered. From his having been exposed for a great length of time to the most inclement weather we have experienced this winter, during the last 13 hours of which he had lain in an [7] insensible state with the sea breaking over him, it appeared clearly to be a case of complete torpor from cold.

The evident indications were, to preserve and encrease the remaining warmth, to endeavour to imitate and restore respiration, and to excite the action of the heart. My first step was, to strip him, which from the rigidity of his limbs, and his being so drenched with water, could only be effected by cutting off his clothes. He was wiped dry, and placed in warm blankets on a bed, with his head and shoulders considerably raised: warm flannels were then applied to the region of the heart, and were frequently renewed. Not being provided with any apparatus for inflating the lungs, my only resource was to depress the ribs, as far as their natural elasticity would allow me, and then suffer them to recover their former position. By repeating this, a quantity of air was alternately expelled from the lungs, and a fresh supply admitted, and that succession thus promoted in the vessels of the bronchiae, which appeared best calculated to imitate the natural respiration. During this time, the assistants were rubbing his limbs, and the other parts of the body, with coarse warm flannels. As soon as water could be warmed, his feet, legs, and hands were immersed in it. His nostrils and mouth were bathed with brandy [8] (the only stimulant I could procure) and some of it was applied to his chest. I endeavoured to get a few teaspoonfuls of warm brandy and water into his stomach, but was some time before I succeeded. After about half an hour's perseverance in these means I perceived a slight heaving of the chest, but it was more than an hour before I could feel any pulse, and then it was extremely weak and irregular. Considerably more than three hours had elapsed, before I felt my hopes of saving my patient at all confirmed: by which time his pulse had gained in strength and regularity and his breathing was more deep and natural. He gradually became comatose, and appeared in a sound sleep, like a person recovering from a fit of epilepsy. From this he was partially roused about ten at night when he muttered a few words, but speedily relapsed into the same state. Next morning he awoke in the possession of his senses and recollection, but greatly wondering at his situation and complaining much of soreness, from the violence which had been used in removing him from the ship. The captain of the vessel has since assured me, he had himself, as well as others, frequently trodden on his body, during the night supposing him dead.

I regret to say I cannot give you the sequel of this case, for my professional avocation obliged me to leave him the following morning, intending to see him the next day. I was greatly surprised however, to find that soon after I left him, he was removed by the positive orders of his commanding officer, [9] though at the imminent hazard of his life. My worthy friend, Mr Maurice, who is ever alive to the calls of humanity, has undertaken to make enquiries concerning him, and will acquaint you with their result.

In addition to what I have already related, it may not be improper to add, that in this case I was forcibly struck with the great difficulty of communicating warmth to

---

[7] That part of the abdomen above the stomach.

the system. For hours after the time of my first seeing my patient, on the removal of the warm applications from any part of his body, the same icy coldness immediately returned: and it was not till respiration was carried on with perfect freedom, and had been long continued, that a general warmth was produced. I am Sir

Yours &c.

William Henchman Crowfoot

[10] At a numerous meeting of gentlemen, held on Tuesday January 7th. 1806 at the King's Head, in Kessingland, to celebrate the success of Mr W.H. Crowfoot in his benevolent and judicious exertions, in restoring to life Serjeant Bubb, of the 28th Regiment after 13 hours of suspended animation through intense cold and to give the reward [of 2 guineas] granted by the Royal Humane Society to John Holmes, John Durrant, Joseph Durrant, and Charles Chipperfield, who with unremitting attention applied the means recommended by Mr Crowfoot, for effecting resuscitation.

Tho: Hunt Esq. in the chair.

It was unanimously resolved:

1   That in consequence of the numerous accidents to which the eastern coast is particularly exposed, it is highly expedient to form a Society on the principle of the Royal Humane Society of London.

2   That the object of this Society shall extend to the application of whatever means may most effectually awaken the exertions of humanity in cases of shipwreck.

3   That the gentlemen present will promote the institution to the utmost of their power, by their own example, and by recommending it to their friends and connections.

4   That the meeting be adjourned to Wednesday the 5th day of February next, to be [11] held at the Queen's Head Inn, in Lowestoft,[8] at 12 o'clock in the forenoon, at which the attendance of every gentleman is solicited who may be desirous to promote the object of the above resolutions.

5   That the thanks of this meeting be given to Dr Hawes, the treasurer of the Royal Humane Society, and to Captain Reed, an Elder Brother of the Trinity House, for the zeal and attention they have bestowed in the case which has given occasion to the plan proposed.

6   That these resolutions be printed in the Norfolk Chronicle, the Yarmouth Herald, the Ipswich Journal, and the Bury papers.

7   That the thanks of this meeting be given to the chairman.

Present   Thos Hunt Esq., chairman.
               William Spurdens, clerk.
               Henry Bence Esq.
               Edward White

---

8   The Queen's Head was in Tyler's Lane (now Compass Street) on the site of a row of nineteenth-century cottages currently used as an annexe by Waveney District Council. There was an assembly room attached to the inn.

Robt Atchinson
John Elph
John Crowfoot
William Aldred
William Cooper
Thomas Cunningham
John Davie
Thomas Meek
John Crowfoot junr.
Thomas Pyman
Edward Baxter
W.H. Crowfoot
Michael Maurice, secretary.

[12] Lowestoft, February 5 1806

At an adjourned meeting of gentlemen assembled this day at the Queen's Head in Lowestoft, in consequence of the success of Mr W.H. Crowfoot in his benevolent exertion for restoring Serjeant Bubb of the 28th Regiment after thirteen hours of apparently suspended animation:
The Revd Bence Bence in the chair. It was unanimously resolved,

1   That in consequence of the numerous accidents to which the eastern coast is particularly exposed, it is highly expedient to institute a Society on the principle of the Royal Humane Society of London and that the gentlemen whose names are subscribed to these resolutions, do constitute such a Society, with the addition of as many others as shall be willing to become members of the same.

2   That this Society shall consist of a president, vice presidents, treasurers, and secretaries: and that the Right Honourable Lord Rous be requested in the name of the Society to become the president. The Right Honourable Lord Huntingfield; Sir Thomas Gooch Bart; and Robert Sparrow Esq. vice presidents; the Revd Bence Bence and the Revd Richard Lockwood, treasurers; and the Revd William Spurdens and the Revd Michael Maurice, secretaries.

3   That this Society shall be distinguished by the name of the Suffolk Humane Society.

[13] 4 That the exertions and the rewards of this Society shall in the first instance, embrace a district including the hundreds of Mutford & Lothingland, Blything and Wangford:9 and that other hundreds be invited to unite themselves with this Society.

5   That this Society shall reward any person who admits a body into his house, under suspended animation, and also all those who shall in any way be instrumental in its recovery.

6   That the object of this Society shall also extend to the application of whatever means may most effectually excite the exertions of humanity, particularly among the peasantry on the coast, in cases of shipwreck.

---

9   That is, the Suffolk coast southwards as far as Thorpe, but not including Aldeburgh.

7   That subscriptions for these purposes be solicited: and that every annual subscriber of ten shillings and sixpence shall be a <u>governor</u>, and shall be entitled to vote in all the concerns of the Society.

8   That any <u>five</u> or more governors in the neighbourhood of the place where an accident shall happen within the views of the Society, shall constitute a <u>committee of reward</u> and shall be empowered to distribute the remunerations of the Society; having previously obtained the concurrence of the president, one of the vice presidents, one of the treasurers, or one of the secretaries.

9   That the medical gentlemen members of this Society, be requested to form themselves into a <u>committee</u> for the purpose of framing such <u>instructions</u> as they shall judge beneficial for promoting the ends of this Society.

[14] 10 That an annual meeting for auditing the accounts of the Society, and soliciting subscriptions, be held alternately at the King's Head in Beccles, the Queen's Head in Lowestoft, the [*blank*] on the 17th day of December, being the anniversary of the day on which Mr W.H. Crowfoot's exertions in the cause of humanity were attended with such signal success.

11   That the bankers in Yarmouth, and in the county of Suffolk be requested to receive subscriptions in aid of the purposes of the Society.

12   That a deputation be appointed to confer with the president and the vice president elect, and other gentlemen assembled at the adjourned Quarter Sessions at Beccles on Saturday next, to request their subscriptions and advice for furthering the views of the Society: and that a meeting be then appointed by public advertisement for framing laws and regulations, and giving a permanent form to the Institution.

13   That minute books be procured by the Society, for recording the minutes of the several meetings of the Society, &c.

Bence Bence, Chairman

14   That the thanks of this Society be given to the chairman for his conduct in the chair.

W. Spurdens
M. Maurice secretaries.

[15] Beccles March 3ᵈ 1806

At an adjourned meeting of the Society, appointed by public advertisement, held this day in conformity with the 12ᵗʰ resolution of the last meeting:

Present,

The Revd G. Spurgeon in the chair

The Revd R. Lockwood	T. Blowers Esq.
B. Bence	C. Arnold Esq.
W. Orgill	H. Bence Esq.
– Wood	Captain Hinton
W. Spurdens	Mr Crowfoot, medical assistant
M. Maurice	Mr W.H. Crowfoot, medical assistant
Mr H.S. Davy, medical assistant	

It was further resolved:

1   That the secretaries be requested and authorized to carry on any correspond-

9

ence, at the Society's expence which they shall judge proper for furthering the ends, or completing the form of this Institution.

2   That the secretaries be authorized to print and distribute such documents as they shall judge proper for printing and distribution.

3   That (a subscription having already been made for the purpose of procuring a life boat, to be stationed on this part of the coast; the surplus of which remains in the hands of R. Sparrow Esq. as a fund for rewarding the men who should navigate it) the Revd Mr Spurgeon V.P. and the Revd Mr Orgill be requested to confer with Mr Sparrow respecting a union of the [16] subscription for the said boat with those of this Society; and giving up the said boat to the Society's direction, if the consent of the subscribers thereto can previously be obtained.

4   That the Revd Mr Avarne be requested to undertake the office of treasurer in the hundred of Blything, and that he will solicit some gentleman in his neighbourhood to execute the office of secretary for the said hundred.

5   That although the subscriptions for this institution be requested in the name of annual subscriptions, it shall nevertheless be understood that no subscriber shall be called on for money, except when – and in such proportion as the demands on the Society shall render it necessary; provided always that no subscriber be called on in any one year for a sum exceeding his annual subscription.

6   That the members now present, or any five of them, do constitute a committee, to meet on the 17th inst. at the Queen's Head in Lowestoft at half past ten o'clock in the forenoon, to promote the object for which this present meeting was appointed, as specified in the latter part of the 12th resolution of the last meeting.

7   That the thanks of this Society be given to the Revd M. Maurice, for his exertions in founding the present Institution.

8   That the thanks of this meeting be given to the chairman for his conduct in the chair.

W.T. Spurdens, secretary

[17] March 10th 1806

Memorandum That in conformity with the 4th resolution of the last meeting I wrote to the Revd Mr Avarne, and have to day received his answer accepting the office of treasurer for the hundred of Blything; and signifying that the Revd Mr Badely of Bramfield and vicar of Westleton, has accepted the office of secretary for the said hundred.   W.T. Spurdens, secretary

At an adjourned meeting of the Suffolk Humane Society held at Lowestoft, March 17th 1806, the Revd G. Spurgeon V.P. in the chair. Present, the Revd G. Spurgeon, the Revd R. Lockwood, the Revd B. Bence, G. Arnold Esq, Captain Hinton, Mr Elph, Mr Baxter, and the Revd M. Maurice.

The Revd G. Spurgeon related the substance of the conference held with R. Sparrow Esq. agreeable to the 3rd resolution of the last meeting – from which it appeared that the funds belonging to the life boat and to the Suffolk H.S. are not at present to be considered as consolidated. Letters were read from Shields, Dover and London, in answer to enquiries made, two of which were deemed so interesting that their contents were requested to be sent to R. Sparrow Esq.

A circular letter explaining the nature and rules of the S.H. Society was read and 500 copies agreed to be printed – together with a list of the subscribers to the Institution. The report of the medical [18] assistants was read, and 3000 were ordered to be printed and circulated. A deputation consisting of the Revd G. Spurgeon and Captain Hinton was appointed to confer with Mr Sparrow, on the best means of rendering the life boat useful, and on such other topics as they might deem connected with the views of the society.

Resolved that the thanks of the meeting be given to the chairman for his conduct in the chair.

M. Maurice, secretary

Halesworth September 9th 1806

At a meeting of the Society held this day at the Angel Inn, Halesworth, pursuant to public advertisement:

Present   The Revd B. Bence in the chair
Rt Hon Lord Huntingfield V.P.

Dr Hamilton	Revd W. Spurdens
Revd T. Avarne	Mr Crowfoot M.A.
N. Orgill	Mr Davy M.A.
W. Buckle T	Mr Bevans M.A.
T. Robinson T	Revd S. Baddely S

A letter was read from Dr William Hamilton of Ipswich expressing his wishes for the success of the Suffolk Humane Society, and accompanied by a pamphlet on the means to be adopted in cases of suspended animation.

[19] A letter was also read from the Revd W. Gee of Ipswich, stating the wishes of the subscribers at Ipswich & the neighbourhood to extend the benefits of this Society throughout the eastern coast.

The chairman delivered a message from the Revd Dr Frank,[10] proposing to extend the benefits of the Suffolk Humane Society to the hundred of Plomesgate.[11] It was unanimously resolved,

1   that the thanks of this meeting be given to Dr W. Hamilton for his letter and pamphlet on resuscitation, and that he be elected an honorary member of this Society.

2   that a letter be written to the Revd Mr Gee expressive of the Society's acquiescence in the plan proposed by the subscribers in Ipswich and its vicinity – and of their desire to further its accomplishment by every means in their power.

3   that a letter be written to the Revd Dr Frank, thanking him for the honour he has done this Society in the message communicated by the chairman, but

---

[10]   The Revd Richard Frank DD, of Alderton, who was one of the promoters of the Bawdsey lifeboat.
[11]   That would have included the Suffolk coast southwards as far as the mouth of the Alde.

recommending a distinct society in conjunction with the other hundreds on the coast.

4   that the following rewards be given by this Society namely, one guinea to the occupier of any house into which a body apparently dead is received: half a guinea to the person who takes a body out of the water: and half a guinea to the person who first brings a medical man to the [20] spot: that a reward be also given to all whom the medical assistant considers as having contributed to the recovery of the patient.

5   that a letter be written by the secretaries to the medical men in their respective districts, requesting them to become medical assistants, and soliciting their attendance at Beccles on the 26th inst.; and that any medical man who will engage to promote by his assistance the plans of the Society, shall be considered a member and entitled to vote at all meetings of the Society, without any pecuniary contribution.

6   that a copy of the affidavits relating to Serjeant Bubb's recovery be sent to Lieut. Col. Johnson, of the 28th Regiment of Foot.

Memorandum that the Revd T. Robinson of Southwold accepted the office of treasurer to this Society.

                        S. Baddely
                        W. Spurdens   secretaries

Ipswich Journal 28 November 1807

Thursday se'nnight was launched the Frances Anne Life Boat, built at Lowestoft under the direction of Mr Lukin, of Long Acre, London, the original inventor and patentee of unemmergible boats. The weather was very unfavourable, an incessant and heavy rain falling all the day. From this cause the number of persons assembled was not so great as it would have been, though some gentlemen, animated by the noble wish of promoting the means for saving the lives of their fellow-creatures, came at a considerable distance to witness the success of the undertaking. At 12 o'clock the boat was launched, wind about S.E. and continued encreasing all the time the boat was at sea. After sailing in various directions she reached the North-end of Corton Sand, upon which the sea and surf were very high. The utility of the boat was eminently shewn in turning the whole length upon the sand without shipping any water. When she came off the sand, the plugs were taken up, and the water suffered to rise as high as the air casks, which were lashed within the boat, would allow. She then stretched under a press of sail to Pakefield; the water with which her bottom was filled, did not appear to retard her progress. There were 16 persons in the boat, including some gentlemen who had volunteered their services. Tho' all of them got over to the leeward side, and some of them stood on the gunwale, yet from all their weight, the press of sail and the plugs still open, her side was not depressed, nor did the water within encrease. On her return near the shore, she was by means of buckets completely filled with water, and the intention was, whilst in that state, that she should receive as many persons on board as was possible. On account of the stormyness of the day, no boat could go off from the beach, but 4 more persons from another vessel were taken in. It is calculated she would have carried 50 persons with safety, when quite full of water. In the melancholy cases which are frequently occurring on this coast, there is every reason to conclude, that many lives will now be saved, which would otherwise be lost. The seamen too, will be enabled to render their assistance, on these occasions, with a confidence and security to which they have been unaccustomed. Too great praise cannot be given to Mr Lukin and Mr Lionel Lukin, his son, for their unwearied attention in superintending the building of this boat, and for prolonging their stay at Lowestoft many weeks, for the sole purpose of seeing her finished. The boat has an iron keel, which serves her for ballast, with a contrivance of casks, placed at her bottom, to be filled with water when necessary to encrease her ballast. Other air casks, for the purpose of buoyancy, and to prevent her sinking, though filled with water, are fixed round her inside. She has also projecting gunwales, with concealed air boxes, and cased with cork. Although the safety of a life boat depends very little upon her form, yet this boat is built after a model which the Lowestoft seamen consider to be best adapted to their shore, for rowing, for sailing, for stowage, and for every other useful purpose. The life boat built by Mr Great-head, of Shields, 1800, and for which there was a very liberal subscription, was found on trial to be unfit for this coast, because of the distance from the sands on which the shipwrecks generally happen, and on account of the difficulty of rowing any boat from the shore, because of the strength of the current and surf. Mr Lukin's boat is calculated to overcome all these obstacles. R. Sparrow, Esq. of Worlingham-hall, whose zeal and endeavours to promote the adoption of life boats have been unceasing, availed himself of the casual and temporary residence of Mr Lukin at Lowestoft, to build the present boat. The funds arising from the old subscription, of which Mr. Sparrow was treasurer, have, by this new undertaking, been exhausted. It is therefore incumbent upon the original subscribers to consider by what mode their benevolent intentions may be most effectually accomplished. There is no doubt that a liberal and humane public will be ready to aid their efforts, and that an appeal will be no sooner made to their countrymen, than their exertions will be shewn in supplying those funds which may provide for current expences, unavoidable accidents, and contribute to preserve the lives of our fellow-creatures, when most exposed to danger.

[21] *[Note in pencil on an otherwise blank sheet]*
  Thus far this appears to be a duplicate of the 1st. Minute Book.

  Here follows a continuation of the Minutes from the 1st. Vol. commencing 31 July 1833 to present time Geo. S. Oct 1 56[12]

[22] At a meeting of a committee of the Suffolk
  Humane Society held 31st July 1833

---

[12] This could be George Seppings, a member of the committee in the 1850s.

Present   John Elph   Lieut. Mathias
                 Robert Reeve[13]   Edmd. Norton[14]
                 Lieut. Carter[15]

At this meeting an estimate for building the life boat house on the Beach having been laid before the meeting by Mr Joseph Bemment,[16] including bricklayer's and carpenter's work and all materials except stones and bricks – all iron work and lock to be included for £71, the work to be began immediately and finished by 20th October next.

Ordered that Mr Bemment's estimate be agreed to and that the work be forthwith begun and proceeded with under the direction of the life boat committee

Edmund Norton, secretary

The Anniversary Meeting of the Suffolk Humane Society held at the Queen's Head Inn Lowestoft on Friday the 6th day of September 1833.
Sir Thos Sherlock Gooch Bart[17] president in the chair.

At this meeting the accounts for the past year were audited and allowed and the balance found to be in the treasurer's hands £158 9s. 11d.

Ordered that the printed instructions for restoration of drowned persons be pasted on boards and distributed at houses near the Beach.[18]

[23] The case of assistance rendered by Samuel Lincoln and Samuel West in saving the life of James Stevens, master of the *Maria* sloop of Ipswich on 16th April last, a statement of which is at foot, having been taken into consideration, Ordered that they be rewarded with the sum of half a guinea.

The case of assistance rendered by George Marshall in saving William Bartlett from drowning on 24th October last, a statement of which is at foot, having been taken into consideration, Ordered that he be rewarded with the sum of half a guinea.[19]

The case of assistance rendered by Charles Triggs and William Wright in rescuing the child of James Balls of Lowestoft, stone mason, from drowning, she having fallen over the groin at the harbour's mouth into the sea and being when rescued sinking in deep water, having been taken into consideration, Ordered that they be rewarded with the sum of half a guinea.

At this meeting the Revd W.R. Brown of Gisleham and Lieut Savage R.N. of Corton were proposed and elected members and notified their intention of becoming annual subscribers of 10s. 6d. each.

At this meeting the life boat committee was reappointed

---

13 Attorney and lord of the manor of Lowestoft.
14 Attorney, superintendent registrar and magistrates' clerk, and secretary of the Suffolk Humane Society.
15 Lieutenant Samuel Thomas Carter, RN, of London Road, Lowestoft.
16 Joseph Bemment, joiner etc., of Old Market, Lowestoft (William White, *History, Gazetteer and Directory of Suffolk* (Sheffield, 1844; reprinted Newton Abbot, 1970), p.510).
17 Of Benacre Hall.
18 The Beach was the name used at Lowestoft for the fishermen's and beachmen's settlement below the cliff which began to be inhabited in the early years of the nineteenth century. From about 1950 it was termed the Beach Village.
19 Ten shillings and sixpence.

Resolved that an application be made to the lord of the manor of Lowestoft for a grant of land on the Beach,[20] on which the new life boat house is building.

The president having received and laid before this meeting a letter from Captain Manby[21] with a report and correspondence as to his apparatus for getting a boat off the beach, the same was ordered to be deposited with the records of this Society.

[Copy statement No.1.]

'April 16th 1833

'At 5 a.m. as James Stephens[22] Master of the *Brittania* (*sic*) [24] sloop belonging to Ipswich was endeavouring to get on board his vessel by himself in his small boat while crossing the Flat off Pakefield he was seized with cramp in his left hand which was the cause of his boat upsetting. He made several attempts to get to the beach by swimming but could not succeed till two men belonging to Pakefield, Samuel Lincoln and Samuel West, ran into the water and brought him out. He is of opinion had not the two before-mentioned men been at hand he should have been lost.'

Witness his hand this 17th day of April 1833

(signed) James Stevens

[Copy statement No.2]

'John Barttell snr, John Barttell junr and William Barttell of Ramsgate, fishermen, were in Lowestoft roads on board a smack called the *Defiance* belonging to Plymouth[23] came on shore in the North Roads at 3 o'clock in the afternoon and were returning on board in the evening when their boat filled with water. In their second endeavour to put off from the shore the boat capsized and all three of the men thrown into the surf at the time very heavy and William Barttell under the boat. He was extricated from his perilous situation by George Marshall who was the only person near them when the accident happened. It being impossible for them to get to their vessel that night George Marshall took them home to his cottage in the Hempland,[24] made them a good fire, dried their clothes and gave them such refreshment as his cottage would afford.'

'William Barttell verily believes that he should have lost his life had he not received the assistance [25] of Marshall.'

'This transaction occurred on 24th October last.'

Edmund Norton

	At a meeting of the committee of the Suffolk Humane Society held 25th October 1833	
Present	Robt. Reeve	Richard Miller
	John Elph	Edmund Norton

---

20 The Beach, or the Denes, was long-standing manorial waste, still in possession of the lord. Before the construction of the harbour it served as an open-air wharf as well as providing rough grazing and net-drying facilities.

21 Captain George Manby, of Yarmouth, inventor of the line-throwing mortar and other lifesaving equipment.

22 He is both Stephens and Stevens in the original.

23 This appears to be evidence of early trawling operations at Lowestoft.

24 A street in the north end of town to the west of High Street, so named because hemp was once grown there for use in ropemaking and the production of linen fibre.

Two of the crew belonging to the *Rebecca* of Burlington[25] wrecked a few days since on the Galloper[26] having lost all their clothes and received no wages – no part of the wreck being saved, applied for assistance to enable them to proceed to Yarmouth on their way home. Ordered that the treasurer pay them 4s. for that purpose out of the Shipwreck Fund.

Edmund Norton

At a Meeting of the committee of the Suffolk
Humane Society held 28th October 1833

Present   Robert Reeve   George Sead Gowing[27]
Revd B. Ritson[28]   Revd F. Cunningham[29]
John Elph

It being represented by Richard Young, master of the *James and Margaret* of Newcastle, laden with coals to Scheedam [*recte* Schiedam], that the vessel foundered on the south east part of the Well Bank on the 23rd instant and that he **[26]** and the crew consisting of five men besides himself after remaining with the wreck two days and two nights reached in the long boat a Norwegian brig bound to Rochell who put them on board the *Betsy* (hosting to Lowestoft), James Burgess master, and that there [*they?*] were set on shore here this morning – and that having lost all their clothes and bedding they petitioned for some relief to enable them to reach their home.

Resolved that they be allowed 10s. each out of the Shipwreck Fund to be paid by the treasurer.

Edmund Norton secretary

At a meeting of the committee of the Suffolk
Humane Society held 29th January 1834

Present   John Francis Leathes Esq.   John Elph Esq.
Revd B. Ritson   Richard Miller Esq.
Thomas Wood Esq.   William Everitt[30]
Revd E.M. Love   Edmund Norton

Richard Consett, master of the brig *Gratitude* of Sunderland, bound to London,

---

25 Bridlington, Yorkshire.

26 A shoal off the Essex coast to the southward of the Inner and Outer Gabbard.

27 Fish curer and boatowner, agent to Lloyd's, High Street, Lowestoft.

28 Perpetual curate of Hopton and for thirty-seven years resident curate of Lowestoft, and later secretary of the Suffolk Humane Society. He died in 1835 at the age of sixty-seven, having been taken ill while preaching at Hopton, and was buried in St Margaret's churchyard at Lowestoft.

29 Francis Cunningham was rector of Pakefield from 1814 to 1855 and vicar of Lowestoft from 1830 to 1860. He married Richenda Gurney, sister of the prison reformer Elizabeth Fry, in 1816. Canon B.P.W. Stather Hunt says in his *Pakefield: The Church and Village* (Lowestoft, 5th edition, 1938), p. 59, that he and his wife 'during the time of their residence kept open house to the whole parish. It is said that it was literally true that the Rectory door (the "Old Rectory" in Pakefield Street) was never shut day or night during the sixteen years in which he lived in the parish.'

30 George and William Everitt, fish curers and boat owners (White, *History, Gazetteer and Directory*, p.509).

applied on behalf of his crew for some assistance to enable them to proceed to their homes. He states that 'Last evening about nine we drove upon the brig *Friends* of Scarboro' – our ship was stove, she filled and was in danger of going down with himself and crew consisting of James Smith mate, Robert Foster carpenter, Michael Robinson, John Whittaker seaman, Noble Hodgson cook, William Coverdale boy and one apprentice, The *Friends* received damage and was run ashore at Pakefield – the men **[27]** saved nothing but what they stood in and believed the men had no money – the vessel is sunk and nothing is saved from her to assist the seamen in getting home.'

Ordered that the five men and boy[31] be relieved with £3 out of the Shipwreck Fund to enable them to proceed towards their homes.

At a meeting of the committee of the Suffolk
Humane Society held 5th July 1834

Present   John Francis Leathes Esq. in the chair

Lieut. Col. Jones	Lieut. Evison
Thos Wood Esq.	Edmund Norton
Revd B. Ritson	Robt Reeve
Lieut. Carter	

At this meeting the treasurer presented the bill for building the life boat shed, and the committee having reported that the building was properly done.

    Mr Joseph Bemment is ordered to be paid.

[28]   The anniversary meeting of the Suffolk
Humane Society held at the Crown Inn Lowestoft
on Friday the 29th day of August 1834
Sir Thomas Sherlock Gooch Bart president in the chair.

At this meeting the treasurer's accounts for the past year were audited and allowed and the balance in the treasurer's hands found to be £116 19s. 4d.

    At this meeting Robert Fowler Esq.[32] was admitted a subscribing member.

    The statement of Benjamin Butcher and others for rescuing the life of a female servant of Mr Seaman's who attempted to drown herself on the sea shore being taken into consideration they were ordered to be rewarded with half a guinea.

    Ordered that William Peek and William Smith be paid half a guinea each for saving the lives of Samuel Mewse and William Boyce who with William Galvin whose life was lost were upset in their boat off Corton on the 28th instant.

Ordered that John Cook, Robert Rounce and Samuel Outlaw be paid Seven shillings and sixpence for their assistance in saving the life of a man who fell overboard from the brig *Ross* of Pillshead[33] on 21st April 1833

---

[31] Master and five men and a boy, and an apprentice: that adds up to eight, not six!
[32] Robert Cooke Fowler Esq., of New Hall, Gunton.
[33] Presumably Pillhead, near Bideford, Devon.

Ordered that Joseph Boyce [blank] and John Cooper for saving the life of George Gale, a boy who was knocked into the harbour, be paid the said John Cooper 5/- and the said Joseph Boyce 2/6d.

At this meeting the life boat committee was reappointed.

Lieut Carter reported that the boat & material were in good repair.

Ordered that a list of subscribers be printed.

[29] The Anniversary Meeting of the Suffolk Humane Society held at the Queen's Head Inn Lowestoft on Wednesday the 18th day of September 1835

Sir Thomas S. Gooch Bart president in the chair.

At this meeting the accounts for the past year were audited and allowed and the balance found to be in the treasurer's hands £136 17 3.

At this meeting Charles Porcher Esquire was unanimously elected a vice president in the room of the Earl of Stradbroke resigned and William Adair Esquire in the room of Alexander Adair Esquire deceased.

At this meeting the Revd [blank] Clissold and the Revd Thomas Sheriffe were admitted subscribing members.

John Cook Junr, James Cook and John Cone applied to the Society for reward for having saved the lives of Richard Tripp, William Garner and Christopher Johnson who were upset in a boat in the North Roads near the Battery[34] in a heavy surf and due enquiry having been made into the facts they were deemed entitled to reward for meritorious services and ordered to be paid half a guinea each.

Mrs Cook applied for her trouble in taking care and nursing Mr Seaman's servant and was ordered to be paid 5/-.

John Hunting, [blank] Charles Goodwin, Thomas Colby, Samuel Martin and William Adams stated that they saw a boat off Pakefield at about 50 yards from the beach with six men in her on the 31st May, that she upset and only four men then appeared. That Hunting swam with a rope round him and brought one man ashore and Thompson and another brought a man named Smith ashore without a rope, and that the other men were assisting **[30]** them on the beach and the facts having been investigated

It is ordered that Thompson and Hunting be rewarded with 10/6d. each and the other four men with five shillings each.

Ordered that John Melchman be paid for brandy and blankets &c. for the parties in the above case 5/-.

John Clarke, a preventive man applied for reward for saving a person overboard in the harbour about two months since. Ordered that he be rewarded with 2/6d.

The following Gentlemen were appointed as a special committee for revising the rules of the Society.

Sir Thomas S. Gooch President

| Charles Porcher Esquire | Chas Steward |
| Robert Reeve | S.T. Carter |

---

[34] The East Battery near Ness Point, established during the Napoleonic Wars. The Ness, the most easterly point of Britain, is about three-quarters of a mile north of the harbour.

Edmd Norton   Revd E. Thurlow
John Elph   Geo Everitt with a quorum of three.

At a meeting of the committee of the Suffolk
Humane Society held 5th November 1835.
President Robert Reeve Esqre. in the chair
John Elph   William Everitt
Samuel T. Carter   G.S. Gowing
Edmund Norton

At this meeting John Wood the master of the *Prince Frederick* steamer bound from Goole to London with a general cargo and wrecked on Corton Sand about five o'clock in the **[31]** morning of Wednesday the 4th instant made application to the committee for assistance for the crew 13 in number to leave Lowestoft towards their respective homes who saved nothing but what they stand in.

Ordered that they be paid £3 10 0 to assist them on the way.
The following are the names of the crew

Thomas Wood   John Elliott
John Brownbrige   John Thompson
William Briggs   Joseph Rufling
John Teather   Joseph Yumerfale
John Singleton cook   Robert Thompson
Robert Whomsley   Thomas Dummion
Joseph Smithes

At a meeting of the committee of the Suffolk
Humane Society held 6th February 1836

Present   Robert Reeve Esq.
Revd Francis Cunningham   S.T. Carter R.N.
Mr George Everitt   Mr G.S. Gowing
Edmund Norton

At this meeting an application was made for the crew of the *David Ricardo* of London consisting of six men (exclusive of master) who were brought on shore at Kessingland by the Lowestoft life boat on Thursday last. And Mr G.S. Gowing and Mr Johnson reported that they had scarcely any clothing to cover them.

Ordered that Mr G.S. Gowing & Mr Johnson be authorized to spend a sum not exceeding £6 towards clothing the men and getting them off towards home.

[32] At this meeting an application was also made for the crew of the *Louisa* of Poole brought on shore from the wreck by lines by the Pakefield boatmen – this crew consists of four men exclusive of the master – and they were represented as having no clothes.

Ordered that Mr G.S. Gowing & Mr Johnson be authorized to expend a sum not exceeding £4 towards clothing the men and getting them off towards home.

The life boat having gone out on Thursday to the assistance of the crews of the wrecked vessels on Thursday last from Lowestoft when she saved one man from the *Speedwell* of Shields and having afterwards been launched from Kessingland beach to the assistance of the *David Ricardo* from which seven men were saved It is ordered that the two crews be paid the usual allowance of £5 each.

A special meeting of the Suffolk Humane
Society holden at the Queen's Head Inn Lowestoft on
Tuesday the 12th day of April 1836 at 12 o'clock

Sir Thomas Gooch Bart president in the chair

Charles Porcher Esq.   Revd Richard Pearson
Revd F. Cunningham   Thos Glasspoole Esq.
Robt Fiske Esq.   Robt. Reeve Esq.
S.T. Carter Lt. RN.   Joseph Chapman

The wife of James Durrant and the wife of Thomas Durrant both of Kessingland applied to the Society for some remuneration for the services rendered by them to William Berbrick saved from the wreck of the *Speedwell* on Kessingland [33] beach on 4th February last, the one in rubbing and nursing him and the other in fetching necessaries from the village.

The meeting having taken same into consideration and made enquiries as to the facts, it was resolved that the wife of James Durrant be paid 8/- and the wife of Thos Durrant 5/-.

The bill of Charles Marjoram for [*blank*] ten men
to Lowestoft belonging to life boat   6.-
William Solomon bill of   10.3

The bills of Thos Cunningham[35] and Robt Manthorpe[36] referred to next Annual Meeting.

Ordered that the Medal of the Society be presented to Mr Joseph Mott Hodgson [*recte* Hodgkin], Surgeon of Kessingland, for the services rendered by him on the coast on the 4th of February last and that he be appointed one of the medical assistants of the Society.

Ipswich Journal 4 February 1836
Lowestoft, Feb. 4.– Early this morning it was blowing a gale from E.N.E. with very heavy seas. Two brigs were observed on the Barnards about 5 miles south from the port. Lieut. T.S. Carter R.N. assisted by many people from the town soon succeeded in launching the lifeboat and with a crew of brave fellows proceeded to the wrecks, one of which, sad to relate, was entirely gone with all her crew. From the other, the Speedwell of South Shields, they succeeded in saving the only man left. This poor fellow they landed at Kessingland. After landing the solitary survivor from the Speedwell at Kessingland it became necessary to convey the lifeboat about ¾ mile to the northward in order to launch to the assistance of the David Ricardo of London, which was then lying on the outer flat. In effecting this object Lieut. Carter unfortunately received severe injury to his right arm which prevented him going off in the

---

35 Wheelwright and victualler, the King's Head, Kessingland.
36 Joiner and beerhouse keeper, Kessingland.

boat. However, she was soon launched, manned by a crew from Lowestoft and Pakefield, who were successful in saving the whole crew of 7. The brig was lying too far out to allow Lieut. Jones to use his apparatus.[37]

The Anniversary Meeting of the Suffolk
Humane Society held at the Crown Inn, Lowestoft,
on Friday the 13th day of September 1836

Sir Thomas S. Gooch Bart In the chair

At this meeting the accounts for the past year were audited and allowed and the balance found to be in the treasurer's hands £179 10 1.

The Revd Edwin Proctor Dennis and Captain Thomas Leggett admitted subscribing members.

At this meeting the Medal of the Society ordered to be presented to Mr Joseph F. Hodgkin at the special meeting of [34] the 12th April last was presented to him.

Samuel Wright of Kessingland having been recommended to the Society by Mr Hodgkin as rendering to him as surgeon very useful assistance [*blank*]

The circumstance of Lieut Carter having met with an accident on the beach at Kessingland on the fourth of February last whilst drawing the life boat along the beach by the breaking of the rope by which his elbow was dislocated having been certified by Mr Hodgkin and various services rendered by Lieut Carter being taken into consideration, it was resolved that the expences incurred by surgeons' bills in consequence of the accident be paid out of the funds of this Society and that the president be requested to make a further application in the name of the Society to the Admiralty for promotion and also to the National Institution for some further reward for the various services he has rendered in saving the lives of shipwrecked mariners upon this coast.

William Bobbitt attended and stated that about the latter end of October the *Sea* of Stockton struck on the Barnard and afterwards came to the beach when the crew and some of the longshore men aboard came ashore at the Ness and was upset into the wash with thirteen men in her, that he and the following men, vizt Benjamin Day, Benjamin Butcher, James Tripp, William Carsey, Benjamin Barnafield, and John Harvey went to their assistance and reached her by taking hold of each other through the surf and took three men and a boy out of the water. It is ordered that the men be rewarded with 30/-.

Lieut. Carter states May 5th James Ayers Junr was the means of saving a man from the brig *Elizabeth* who when coming to the shore for the master the boat upset in the breakers at the harbour's mouth and Ayers ran into the [35] water and saved one of the persons. Ordered that Ayers be rewarded with 5/-.

The bill of Thomas Cunningham[38] for refreshments &c. for the life boat crew on 4th February 1836 having been presented it was ordered that one pound only be allowed of the bill the same having been incurred without any orders.

---

[37] The line-throwing mortar apparatus.
[38] Licensee of the King's Head, Kessingland, and wheelwright.

The following gentlemen were appointed to form the life boat committee for the ensuing year, viz. the treasurer and secretary, Mr G. Everitt, Mr J.S. Lincoln, Lieut Carter, Mr G.S. Gowing and Mr William Everitt.

My Lords

Whilst it must always be a matter of pride to the British Nation to know that the number of applications for promotion to your Honourable Board by no means equals that of meritorious services performed it is at the same time undoubtedly true that the vast variety of cases (nearly equal perhaps in merit) renders the selection of the most deserving very difficult to your Lordships. It is however hoped by your memorialists that the admission of this truth will not prejudice the object of the present address, than whom we may confidently assert no officer can shew in his peculiar way a more extended list of intrepid services performed. It may moreover be safely, because with truth, asserted that this present document originates with the most influential and respectable portion of the inhabitants of that part of the sea coast of Suffolk to which his exertions have been in this respect [36] confined and where therefore they can be most appreciated.

Lieut Carter RN aged 52 was employed in active service from 1799 to 1814 inclusive, of which ten years were in the West Indies and the rest in the Channel, North Sea and North America. Since the Peace he has been during the last eighteen years resident at Lowestoft and during all that time has been honorary commander of the life boat belonging to the Humane Society of that place; on every occasion on which the boat could possibly be used for the preservation of the lives of ship-wrecked mariners on this dangerous coast Lieut Carter has always been ready and has always taken the command of her. He has thus been the instrument under God of rescuing no less than seventy two lives (see Schedule) from impending destruction.

On several of these occasions this gallant officer has recklessly exposed his own life to imminent peril and hazard to save that of his fellow creatures and only on a late occasion on the 13th September last his anxiety to bring the life boat into service was the cause of his having his elbow dislocated and his suffering other bodily injury from which he is still to some extent affected.

Lieut. Carter's exertions have not been alone confined to the active duties of the boat when on service, but he has been for seventeen years indefatigable in his care as to her effective state as well as to her repairs, as to the safe and proper state of all her fittings, tackle &c. and has kept them at all times in so efficient a state as to have called forth on all occasions the warm praise of naval officers visiting the town of Lowestoft and these exertions so essential to the real usefulness of the boat will be the more appreciated from their continuity and as being proofs of his seaman like qualities.

The Suffolk Humane Society have done all they could to testify their warmest approbation of his integrity and zeal. [37] Thanks are given to him as often as the Society meets and the only honorary distinction they can present, their Medal, has been long since given to him. It is because they feel themselves unable adequately to reward such services that they resolved at their last annual meeting to address your Lordships to extend your patronage to this useful officer by giving him promotion, and they and the undersigned persons confidently hope that such services as those peculiarly the object of this memorial will have great weight with your Lordships as well from the failure of other employment for officers of the Navy as from the test given above of Lieut. Carter's merits and energy and because the signatures attached hereto will forbid you to think that any interested feeling can actuate those who have subscribed them.

To the Lords Commissioners
of the Admiralty &c. &c.

Lowestoft
8th Decr. 1836

Signed

T.S. Gooch president
Charles Blois VP, JP
J.F. Leathes VP, JP
Charles Porcher VP, JP
William Jones JP for Suffolk
Edw Missenden Love JP for Suffolk
Henry Bence Bence JP
Dawson Turner JP[39]
Edw Montagu JP
John Brightwen
Robt Reeve, treasurer SHS

Edmund Norton Secy SHS
Geo Edwards, Lowestoft Harbour
Francis Cunningham, vicar of Lowestoft
E.C. Sharpin, solicitor, Beccles
William Cleveland
G.S. Gowing
John Elph
James Everard
Charles W. Smith
J.W. Hickling
G. Everitt

[38] To the Governors of the National
Shipwreck Institution

Lieut. Samuel Thomas Carter RN has been a resident at Lowestoft upwards of seventeen years, and during the whole of that time has been the honorary commander of the life boat, belonging to the Humane Society of that place. On every occasion on which the boat could be possibly used for the preservation of the lives of shipwrecked mariners on this dangerous and difficult coast, this gallant officer has been ready and has always taken the command of her – He has thus been the instrument, under Providence of rescuing no less than eighty four persons[40] from impending destruction (See Schedule).

On several of these occasions Lieut Carter has recklessly exposed his own life to imminent peril and hazard, to save that of his fellow creatures and on a late occasion on the 13th of September last his anxiety to bring the life boat into service was the cause of his having his elbow dislocated and his suffering other bodily injury from which he is still to a great degree affected.

Lieut. Carter's exertions have not been confined alone to the active service of the boat when at sea but he has from the time he took the command been indefatigable in his unremitting care as to her effective state as well as to her repairs, as to the safe and proper state of all her fittings, tackle &c. And has kept them at all times in so efficient a state as to have called forth on all occasions the warm praise of

---

39 Dawson Turner (1775–1858) was a Yarmouth banker, a noted botanist and collaborator of his son-in-law William Jackson Hooker, and an influential antiquarian. His mother was related to the Norwich School artist John Sell Cotman, who assisted him with botanical illustrations and with whom he produced several illustrated books. Cotman taught Turner's daughters to draw and paint.
40 Twelve more than mentioned in the letter to their Lordships of the Admiralty; this letter must have been sent after the rescue of twelve men from the *Prince of Brazil* on 14 January 1837.

naval officers visiting this watering place and these exertions so essential to the real usefulness of the boat will be more appreciated from their continuity and as being proofs of his seamanlike [39] qualities.

The Suffolk Humane Society have done (however little) all they could to testify their warmest approbation of his noble conduct, the thanks of the Society are constantly given to him and the only honorary distinction they can present, or he would accept, their Medal, has been long since given to him.

It is because they feel themselves unable properly or adequately to reward such exertions that they thus venture to address your Society, requesting you to extend your patronage to this useful officer by giving him an award of distinction, confidently hoping that such services as those recited above will have great weight with your committee and entitle Lieut. Carter to the consideration of the National Shipwreck Institution

Lowestoft
December 1836

Signed.

T.S. Gooch president
Charles Blois VP, JP
J.F. Leathes VP, JP
Charles Porcher VP, JP
William Jones JP for Suffolk
Edward Missenden Love JP for Suffolk
Dawson Turner JP
John Brightwen

Robert Reeve, treasurer SHS
Edmund Norton, seretary SHS
Geo Edwards, Lowestoft Harbour
E.C. Sharpin, solicitor, Beccles
Henry Bence Bence JP
William Cleveland
G.S. Gowing
James Everard

Charles W. Smith
J.W. Hickling
Francis Cunningham, vicar of Lowestoft
G. Everitt

[40]

A schedule of persons saved by Lowestoft life boat
under the command of Lieut. S.T. Carter

Date	Vessels	Masters	Number saved
22d October 1821	Brig *George* of London	John Dixon	7
same day	Sloop *Sarah & Caroline* of Woodbridge	Joseph Spalding	5
18th January 1825	Sloop **Dorset* of London	Hart	2
3d April 1828	Schooner *Suck's all* of London	Joseph Gibson	4
17th May	Brig **Fawn* of Sunderland		10
3d December	Brig **Ann* of Sunderland	Walker	6
23d November 1829	Brig **Thomas & Mary* of Newcastle	Wilson	10
24th	Brig *Ann* of London		9
21st November 1832	Brig *Costenside* of Newcastle	Thomas Pearson	11
21st February 1833	Brig *Tay* of Bridlington	John Farnell	5
19th January 1835	Schooner *Bishop Blaize*		2
6th February 1836	Brig *Speedwell* of Shields		1
14th January 1837	Brig *Prince of Brazil*	David Forward	12
	In the cases marked thus (*) Lieut. Carter		84

24

was assisted by Lieut S.F. Harmer RN
as second in command.

Lieut. Carter's age 52. Five children.

In active service from 1799 to 1814 inclusive of which
ten years in West Indies and the residue in the
Channel and North Sea and North America.

[41]   The Anniversary Meeting of the Suffolk
Humane Society held at the Queen's Head Inn
Lowestoft on Friday the 22nd day of September 1837

Sir Thomas S. Gooch Bart   In the chair

At this meeting the accounts for the past year were audited and allowed and the
balance found to be in the treasurer's hands £195 12 11.

Mr Charles Hursthouse and Mr William V. Barnard were elected subscribing
members at half a guinea per annum and Revd F. Cubitt of Fritton at [blank]

John Rose[41] applied to the Society for a reward for saving the life of Joseph
Fletcher's[42] child about 12 years old who was got out of his depth at the North
Ness about three months since – states that he went in up to his throat and took out
the child. The child was black and blood ran out of his mouth. Ordered that he be
rewarded with 7s/6d.

William Hall applied for a reward for having on the 17th September instant saved
the life of James Patterson who fell into Bobbits Dyke on the Beach. Ordered that
he be rewarded with 2s/6d.

William Crow applied for a reward for saving the life of George Williams who
fell into Robert Tripps pulk-hole[43] on 8th March last – states he was called by a
woman to pick him out which he did the child was about five years old and was
nearly dead – allowed 4s/-.

James Balls applied for a reward for saving Christopher Gilby.

Ordered that he be rewarded with a £1.

[42] The draft of the rules of this Society were submitted to the meeting and approved
– It was ordered that the new rules be printed and circulated with a list of members.

Mr George Cox attended to represent the case of   [blank]
[blank line]
Ordered that they be rewarded with a £1.

Edward Elliss applied for saving a man who fell overboard – he was in a vessel
and laid hold of his wrist – allowed 2s/6d.

---

[41]   A forebear of Jack Rose, the popular Lowestoft historian and lifeboatman.

[42]   'Posh' Fletcher, the protégé of Edward Fitzgerald. 'Old Fitz' had built for him a fishing boat named
Meum et Tuum: it was known at Lowestoft as the *Mum Tum*.

[43]   A small pond, especially a very muddy one. There were several of these on the Denes at one time,
acting as drainage sumps and sewage pits.

At a meeting of the committee of the Suffolk Humane Society held at Mr Norton's office on 3rd November 1837.
Present Charles Porcher Esq. V.P.

Revd F. Cunningham

Lieut S.T. Carter

Samuel Johnson

W.C. Worthington

G.S. Gowing

Revd R.A. Arnold

At this meeting Mr Benjamin Wells Browne was admitted a member at annual subscription of 10s/6d.

Mr William Banfield, master of the brig *Bywell* of Newcastle, 244 tons, lost on the south west end of the Newcome on Wednesday evening the 1st November, applied for some means of sending off the crew towards their home – mate George Palmer, Robert Forkes, carpenter, William Dining, John McLeod, George Esperington, John Carver, John James, William Forgill, lad, Alexander Mijoueth, boy, and William **[43]** Hopwell, an apprentice, saving only the clothes they stand in.

Ordered that the crew be paid £2 to assist them in proceeding on their way home and that George Esperington be allowed 5/- for a pair of shoes.

Suffolk Chronicle 4 November 1836

Lowestoft, Nov. 2. – Last night abut six o'clock P.M. it blew a heavy gale from the South West, with storms of hail and rain. A fine new brig, the Bywell, a Newcastle Trader, 250 tons burthen, William Binfield Master, from London to Newcastle with a general cargo, struck on the Newcome Sand – the sea making a free passage over her. They immediately hoisted lights fore and aft, which were promptly answered by the launching of the lifeboat through a heavy surf, with a spirited crew under the command of Lieut. Carter. These brave men succeeded in rescuing the crew, ten in number, with 2 male and 2 female passengers, from a watery grave. The lifeboat had not left the Bywell above a quarter of an hour before she fell on her beam ends and filled with water, where she now lies, her cargo floating about the sea. Thus in a few hours were a fine ship and cargo worth at least £30,000 lost.

At a meeting of the committee of the Suffolk Humane Society held at Mr Norton's office on 16th November 1837.

Present   Robert Reeve Esq. in the chair

Lieut. Carter   Mr Samuel Johnson

Mr J.S. Lincoln   Mr George S. Gowing

John Drew, captain of the *Jubilee* of Plymouth, coal laden, applied for some assistance for his crew consisting of seven men towards returning home, the vessel being wrecked yesterday and sunk in deep water off Pakefield having struck the wreck of the *Bywell*.

This meeting having taken the case into consideration, resolved to allow them ten shillings each out of the fund for assisting shipwrecked seamen and ordered the treasurer to pay the sum viz. £3 10 0.

At a meeting of the committee of the Suffolk Humane Society held at the Crown Inn on Wednesday the 7th day of February 1838
[44] Present John F. Leathes Esq. V.P.

| Revd E.M. Love | Mr George Everitt |
| Lieut. S.T. Carter R.N. | Mr Samuel Johnson |

Edmund Norton secretary

Ordered that the annual sum of £5 be paid to the crew of the lifeboat for their services on going out to the smack *Walter Scott* of Leath [Leith], Drip, master, on the 10th January 1838.

That Mr George Cox be paid £3 18 for getting the boat home from Kessingland beach on the above occasion.

That [blank] Hales and [blank] Mantripp be paid 5/- each for watching the boat and stores all night.

That Mr John Goldsmith's bill for horses on the above occasion £2 15 be paid. And Robert Fox for allowance to crew and horse drivers 7/-.

At this meeting Mr Hole was proposed and elected a member at 10/6 per annum.

Norwich Mercury 27 January 1838
Lowestoft, Jan. 25
On the 10th inst. a Leith smack was observed at anchor, close under the Barnard Sand, having a signal of distress flying. At ten o'clock the life boat, under the command of Lieut. Carter, was launched, and proceeding to the vessel, which proved to be the Sir Walter Scott, of and from Leith, bound to London. She had lost her rudder and was making a great deal of water; the wind at the time was blowing hard at ENE accompanied with snow. On finding the vessel in no imminent danger, and that the captain and crew were not disposed to leave her, Lieut. C. made a signal to the Seaflower yawl to remain by her, and taking out a female and a boy they landed at Benacre, where the life-boat was left, at the request of the Captain, in the event of her assistance being required.

The Anniversary Meeting of the Suffolk Humane Society holden at the Crown Inn, Lowestoft, on Thursday the 30th day of August 1838.
The president Sir Thomas S. Gooch Bart. in the chair.

At this meeting the treasurer's accounts for the past year were audited and allowed and the balance found to be in the treasurer's hands £200 4 0.

The following members were proposed and admitted [45] members.

Mr Abraham Scales	½ guinea per annum	
Robt Henry Salmon	½	do.
John Hole	½	do.
William Jones Woods	½	do.

At this meeting it was ordered that the sum of £5 be applied for defraying surgeon's bill for attending Lieut. Carter when he dislocated his arm on service with the boat.

Lieut. Carter stated to the meeting that John Davie[44] of Kessingland before daylight on the morning of 10th January last came over express from Kessingland and gave notice of a wrecked vessel which proved to be the *Walter Scott* to which life boat went and from which two persons were saved – and but for which timely information the two men[45] saved must have been lost and Lieut. Carter also reported that the said John Davie has on many occasions been of great service on occasion of wrecks and has always been ready to render assistance when he would be in any way useful.

It is ordered that the Medal of the Society be presented to the said John Davie.

Ordered that 10/- be paid to William Dowson and Samuel Gilby for their assistance in saving the life of Samuel Butcher on 15th May last, he having gone out of his depth whilst going off shrimping.

The case of assistance rendered by Christopher Gilby, James Reynolds, Ham Dowson, and three others on 21st May last to William Burgess, master of the *Rose and Elizabeth*, three men and a boy who were upset in a boat when just off the beach and saved by means of the assistance rendered, having been taken into consideration, ordered that they be rewarded with a £1.

[46] The case of assistance rendered by Benjamin Day who about 5 o'clock in the afternoon of 28th July last saw a child in the water near the Young Company's yawl, and having heard others call out went down to the beach up to about his waist and took the child out of the water – the child was without shoes and stockings and was taken off by the draw of the sea – the child could not speak till it was taken home vizt. Mr Williams of the preventive service[46] aged four years, having been taken into consideration, ordered that he be allowed 2/6d.

The following life boat committee were appointed for the ensuing year, the treasurer and secretary, Charles Porcher Esq., Mr George Everitt, J.S. Lincoln, Lt. Carter, Mr G.S. Gowing, & Mr William Everitt.

Ordered that a bill of George Cox[47] of £2 9 10 be paid but that no bill in future be paid unless it clearly appears to be properly ordered and that items &c. properly shewn.

At this meeting Lieut. Carter and Lieut. Joachim[48] were elected honorary members.

And the thanks of the meeting voted to the chairman.

A committee meeting of the Suffolk Humane Society held at Mr Norton's office on Tuesday the 8th day of January 1839.

Present Edmund Norton secretary
Mr George Everitt   Mr G.S. Gowing
           Mr Samuel Johnson

[47] William Campbell, master of the *George and Henry* of Stockton, ran down and sunk in the North Roads on the 6th instant about 8 in the evening, applied for

---

44 John Davie, farmer, of Kessingland.
45 In fact the two people brought ashore were a woman and a boy.
46 Presumably the boy's father.
47 George Cox, fish curer and boatowner.
48 Lieutenant Richard Joachim, RN, chief officer of coastguard at Lowestoft.

assistance to enable his shipwrecked crew, vizt. Charles Raxby, John Smith mate and James Bell [*to proceed on their way home*] and stated that they escaped the vessel without any clothing but such as they stood in at the time – the mate had his oilskin dress on and was on the watch, the rest of the men were in bed and escaped with little clothing. There were also two apprentices escaped.

Ordered that the crew be relieved with £2 and that Mr Gowing purchase a jacket and cap for the cook.

Edmund Norton, secretary

A committee meeting of the Suffolk Humane Society held at Mr Norton's office on Wednesday the 16th January 1839.
Present J.F. Leathes V.P. in the chair.
Mr George Everitt
– Edmund Norton
– Samuel Johnson
– J.S. Lincoln

William Sutherson, master of the brig *Good Intent* of Newcastle, 281 tons register, states that his vessel (cutaway mainmast) went on her beamsends [*sic*] at sea in the North Seas on the 10th instant; the *Sea* of Hartlepool, Henry Button master, took the crew consisting of the master, mate and seven men whom he brought into Lowestoft Roads this morning – states that he the mate and three men are going down **[48]** to Newcastle in the *Sea* and that he has three men belonging to London, vizt. William Markham, John [*blank*], Charles [*blank*], Ralph [*blank*], and one man belonging to Newcastle named Thomas Dickons, who want some assistance on their way homewards – they have all saved but one suit of clothes which they stand in.

Ordered that the three men bound for London be relieved with 7/6d. each and the Newcastle man with 5/-.

Edmund Norton, secretary

The Anniversary Meeting of the Suffolk Humane Society holden at the Queen's Head Inn, Lowestoft, on Friday the 13th day of September 1839.
Charles Steward Esq. in the chair

At this meeting the treasurer's accounts for the past year were audited and allowed and the balance found to be in the treasurer's hands £ [*blank*]

An application was made on behalf of Robert Golder and William Day for a reward for saving the lives of Daniel Ayers and three of his crew when coming ashore on the morning of 11th May 1839. Ordered that they be rewarded with £1.

An application was made on behalf of John Glansford of Ramsgate for saving the life of Benjamin Taylor of Lowestoft on 9th August **[49]** 1839 – under the following circumstances – Taylor and Outlaw were off in a punt[49] in the Roads about half a mile from the shore – when they were ran down by a French boat – Outlaw jumped

---

49 A small fishing boat operating off the beach.

on board Glansford's fishing boat but Taylor got hold of the keel of the punt – Glansford threw a rope to Taylor and hauled him on board his smack and ran a risk of running his vessel on shore Ordered that he be rewarded with 10/-.

The following persons were proposed and admitted members

Mr J.W. Hickling	10s. 6d.	per annum
S.S. Brame	10s. 6d.	do.
William Searle	10s.	do.
Mrs Gooch	1 1s. 0d.	do.
Mr John Martin	10s. 6d.	do.

At this meeting Captain Harmer[50] was elected an honorary member in consideration of his past services to the Society.

That 6/- for a frock coat lost by William Hales be allowed.

Mr Hole and Mr Brame were elected medical assistants.

Resolved that a special meeting of the Society be called on Wednesday the 9th day of October at eleven o'clock in the forenoon at the Queen's Head to take into consideration the propriety of establishing a life boat at or near Pakefield.

The following life boat committee were appointed for the ensuing year. The treasurer [50] and secretary, Charles Steward Esq., Mr George Everitt, Mr J.S. Lincoln, Lt. Carter, George S. Gowing, Mr William Everitt, and Captain Fowler, Mr J.W. Hickling.

A special meeting of the Suffolk Humane Society held at the Queen's Head Inn, Lowestoft, on Wednesday the 9th day of October 1839 at 11 o'clock – for the purpose of considering the propriety of taking under the controul and management of the Society a second life boat proposed to be built and stationed at Pakefield and to make such order and arrangements relative thereto as may appear expedient.
The president Sir T. F. Gooch Bart in the chair

The object of this meeting having been taken into consideration and discussed and William Colby, Benjamin Tompson, John Lewis, J. Cook, Revd Francis Cullingham[51] and James Rumpf and other persons from Pakefield having attended and it being explained to them that this Society could not go into the question of taking the proposed boat under its controul but upon the distinct understanding that such boat must be the absolute property of the Society and that the boat and its crews and the management thereof must be under the sole and absolute controul, direction and management of this Society upon the same principle in all respects [51] as the present life boat at Lowestoft and particularly that such boat or any part of the crew should never under any circumstances be connected in any respect with matters of salvage, to which terms the Pakefield people declared their acquiescence.

It was moved by Mr George S. Gowing and seconded by Mr George Everitt and unanimously resolved,

That if a sufficient sum of money can be raised for the purpose of building and equipping an efficient life boat to be stationed at or near Pakefield with a proper

---

[50] Captain Samuel Fielding Harmer, RN. He died in China in 1843 while in command of HM steam frigate *Driver*.

[51] Error for Cunningham.

house, that such boat be built and equipped under the direction of a committee of this Society and be taken under the controul and management of this Society in the same manner in all respects as the Lowestoft life boat. The following to be the committee, vizt. the treasurer, and secretary, Messrs George Everitt, J.S. Lincoln, Lt. Carter, G.S. Gowing, William Everitt, Captain Steward, Captain Fowler, J.W. Hickling, S. Johnson, Abraham Scales – five to be a quorum.

A meeting of the committee appointed for superintending the building of a life boat to be stationed at Pakefield. [*no place of meeting or date given*]

[52]

Present   Charles Steward Esq. in the chair

Robert C. Fowler Esq.	Lieut. Carter
Mr George Everitt	Mr J.S. Lincoln
Mr Edmund Norton	” S. Johnson

At this meeting Mr Norton laid before the committee a letter received from Revd H. Cunningham[52] dated 14th February which was read.

It was resolved that this committee are of opinion that they are now in a position to receive to carry into effect the resolution passed at a special meeting of the committee on the 9th October last.

It was determined that the dimensions &c. be taken and further enquiries made as to the opinion of the Pakefield boatmen as to the construction of the boat.

The meeting was then adjourned to Wednesday 11th March next at 12 o'clock.

A meeting of the committee appointed for superintending the building of a life boat to be stationed at Pakefield. [*no place of meeting or date given*]

Present   Chas Steward Esq. in the chair

Mr George Everitt	Pakefield men
Lieut Carter R.N.	Henry Colby
Mr Saml. Johnson	William Colby Junr.
Mr Abraham Scales	Nathaniel Colby
Robert C. Fowler Esq.	Thomas Lewis
	Wm [*blank*][53]
	Samuel Peck

[53] At this meeting the above Pakefield men attended and explained their views of the size and kind of boat they thought best adapted to the Pakefield beach, and they proposed that she should be on the model of their yawl called *The Rescue* and that her length be 44 feet, breadth 12 feet and depth 4 feet with a flat floor. It was determined that the boat be built of English oak.

It was resolved that application be made to Messrs. Barsham[54] and Sparham[55] to

---

52   Probably the Revd Francis Cunningham, vicar of Lowestoft.
53   Probably William Everitt.
54   Error for Bachelor Barcham.
55   Samuel Sparham, boatbuilder on the Beach.

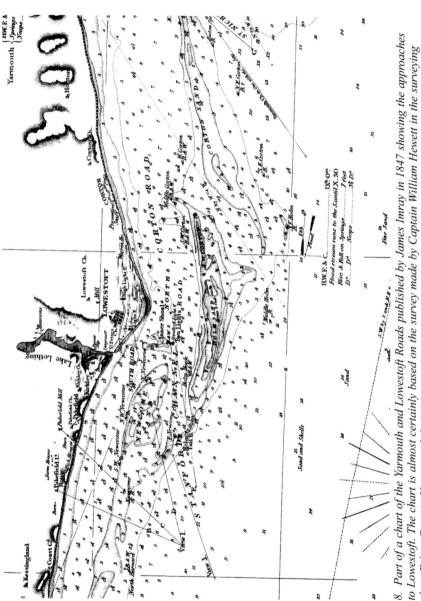

8. Part of a chart of the Yarmouth and Lowestoft Roads published by James Imray in 1847 showing the approaches to Lowestoft. The chart is almost certainly based on the survey made by Captain William Hewett in the surveying sloop Fairy. Captain Hewett and the Fairy were lost south-east of Lowestoft in a severe gale in 1840, but the Hewett Channel off Gorleston preserves his name to the present day.

send in tenders and Mr Teasdall[56] of Yarmouth boatbuilders to send in estimates for the building of the boat, but it is understood that in entering into any contracts the committee will not be bound to accept the lowest tender; resolved that enquiries be made as to the probable cost of the fittings, spars, sails &c.

And that the boat be iron fastened with iron keel, but that the nails and iron be wrought. The nails to be ¾ heads with ½ inch sides.

The meeting adjourned to Wednesday next at 12 o'clock.

A meeting of the committee appointed for superintending the building of the life boat to be stationed at Pakefield holden at the Queen's Head Inn Lowestoft on Wednesday 29th April 1840.

Present Chas Steward Esq. in the chair

Mr George Everitt   Mr Edmund Norton

[54] Mr G.S. Gowing   Mr William Everitt

At this meeting tenders for building the Pakefield life boat were received, vizt. that of Mr William Teasdell of Yarmouth at £177 10s., with a deduction for tanks on the thwarts of £7 10s., and that of Mr Samuel Sparham of Lowestoft at £205, when it appeared adviseable to accept Mr Teasdell's tender subject to some explanations to be given by him on several points in the specification.

And this meeting was adjourned till Wednesday next. And [*last word crossed out*]

A meeting of the committee appointed for superintending the building of the life boat to be stationed at Pakefield held 6th May 1840.

Present Robert C. Fowler Esq. in the chair.

Mr George Everitt   Mr Edmund Norton

Mr G.S. Gowing   Mr James W. Hickling

       and Mr Samuel Johnson

At this meeting Mr Teasdell attended and stated that the difference between the English oak and Quebec oak would be £3 10s. – that the proper proportionate depth of the boat should be 3 feet 8 inches and that the difference between the tender as sent in and making the planks ¾ of an inch from the saw would be £2, making an increase on his estimate of £5 10 – the iron to be [55] English iron home wrought and the iron work to include thowles,[57] bumpkin, tiller, stanchion &c. complete – the boat to have the same number of plugs as the Yarmouth boat, about six of 5 inches bore and plug holes coppered – to have three coats paint, two varnish over one coat of linseed oil. And it being settled that Mr Teasdell would undertake the work with the several alterations suggested at £180, his tender was accepted.

And it was agreed between him and the committee that Captain Harmer R.N. be the mutual referee as to all matters in difference arising under the contract.

---

[56] William Teasdel.

[57] Tholepins.

The Anniversary Meeting of the Suffolk Humane Society holden at the Crown Inn
Lowestoft on Friday the 4th day of September 1840.
Sir T.S. Gooch Bart in the chair.

At this meeting the treasurer's accounts were audited and allowed and the balance
found to be in the treasurer's hands £267 3s. 3d.
Ordered that Mr R.H. Salmon's[58] bill of £4 2s. 5d. for printing be paid.
An application was made on behalf of Christopher Gilby and his boat crews for
saving a man who fell over board out of a ship in the North Roads, which they were
alongside of. Ordered that they be rewarded with S.5.[59]
[56] An application was made on behalf of James Betts for a reward for saving a
child about five years of age, son of John Scowen, from a watery grave Ordered
that he be rewarded with 10/6d.

At this meeting Mr Edmund Norton was appointed treasurer in the room of Robert
Reeve Esq. deceased. And Mr Richard Henry Reeve was appointed secretary in the
room of Mr Norton.
The following persons were proposed and admitted members:

Mr T.F. Reeve	£1 1 0	per annum
Mr R.H. Reeve	10 6	do.
Mr James Cleveland	10 6	
Mr William Lincoln	10 6	
Mr Thomas Preston	10 6	

Ordered that a new list of subscribers be printed and published.

A meeting of the committee appointed for superintending the building of the life
boat to be stationed at Pakefield holden at the Queen's Head Inn 25th November
1840
Present Charles Steward Esq. in the chair.

Robt C. Fowler Esq.	Lieut. Carter
Mr George Everitt	Mr Samuel Johnson
[57] Mr J.S. Lincoln	Mr G.S. Gowing
Mr Edmund Norton	Mr R.H. Reeve

At this meeting it was determined that application be made to Messrs Bemment &
Scarll of Lowestoft and Mr Wright of Pakefield[60] to send in estimates for building a
boat house to be made of wood for the new life boat to be kept in.
It was resolved that the boat house be built in sections and that it be roofed with
Duchess's slate.[61]
It was determined that it would be highly advantagious to remove the Pakefield boat
one score[62] nearer to the southward.

---

[58] Robert Henry Salmon, bookseller and printer (White, *History, Gazetteer and Directory*, p.509).
[59] Possibly five shillings. The equivalent of 25 pence in today's money, but worth a good deal more in 1840.
[60] Joseph Bemmant, joiner, Old Market; William Scarle, cabinet maker; and Michael Wright, baker and joiner, of Pakefield.
[61] Duchess was a size of Welsh slate.
[62] A score is a cleft in the cliff or the sandhills giving access to the beach.

Ordered that the Pakefield men be paid £2 7s. for getting the Pakefield boat home from Yarmouth and other expenses. They are to be allowed ¾ of a cwt. of rope for compensation for the rope spoilt in endeavouring to get the boat off on Saturday last, and that the rope be paid for out of the general fund.

Resolved that the Pakefield & Lowestoft crews be paid 30/- each for their exertions on Saturday last in endeavouring to get the Pakefield life boat off. And this meeting was adjourned to Monday next at ½ past 11.

R.H. Reeve
secretary

A meeting of the committee appointed for superintending the building of the life boat to be stationed at Pakefield holden at the Queen's Head Inn 30th November 1840

[58]   Present Charles Steward Esq. in the chair
          Mr William Everitt   Lieut Carter
          Mr J.S. Lincoln   Mr Edmd Norton
          Mr Samuel Johnson   Mr R.H. Reeve

At this meeting Mr E. Crowe's bill £2 2 4 for towing the Pakefield life boat be allowed. Ordered that a new mast be made by Mr Wright and that Mr Teasdell should exchange the handles of the plugs were [where] necessary and rotted in the handle, and that Mr Wright should place additional gangboards and to have the lower pintle of the rudder repaired and a small chain affixed, and to make a deal tree for the pump.

Ordered that Mr Teasdell send over two pair more oars, one pair 17 feet and one pair 16 feet and that 20 launching timbers,[63] hawser rope and larger belt for the stern be ordered.

The committee inspected the land along the coast and fixed upon a site for the new boat house on the south side of the first piece of land belonging to Captain Leggett. Mr Teasdell delivered his bill amounting to £194 1 5 which was allowed and the treasurer was directed to pay it.

R.H. Reeve
secretary

[59] A meeting of the committee appointed for superintending the building of a life boat to be stationed at Pakefield holden at the Crown Inn 13th January 1841
Present Charles Steward Esq. in the chair
Lieut. Carter   Mr R.H. Reeve
Mr Abraham Scales   Mr William Everitt

At this meeting ordered that the following bills be allowed:

---

63 Presumably what are known locally as skeets, lengths of timber over which the boat is slid down the beach.

59

A Meeting of the Committee appointed for
Superintending the Building of a Life Boat
to be stationed at Pakefield holden at the
Crown Inn 13th January 1841
Present Chas. Steward Esqre in the Chair
Lieut Carter   Mr. RH Reeve
Mr. Abraham Scales   „ William Everitt
At this Meeting ordered that the following
bills be allowed
                                    £   1   0
Abraham Scales   7. 6. 1
Saml Johnson   5. 17. 1
William Allen   3. 4. 6
Michael Wright   10. 18. 6
Henry Colby and others   3. 1. 7
William Colby   „ 18. 6
William Balls   „ 1. 6
    „   „   1. 0. 9
C & B Tilmouth   11. 17. 1

The above bills having been allowed
ordered that the Treasurer discharge them

                              RH Reeve
                                  Secretary

A Meeting of the Committee appointed
for superintending the building of a Life
Boat to be stationed at Pakefield holden at
the Crown Inn Lowestoft 3rd March 1841
Mr. Geo Everitt in the Chair
Lieut Carter RN.   Mr. G. S. Gowing
Mr. Wm. Everitt   „ Edmd. Norton
and RH Reeve Secretary
Ordered that Mr. G. S. Gowings bills of

9. *Page 59 from the Minute Book for 13 January 1841, courtesy of the Lowestoft
Fishermen's and Seafarers' Benevolent Society.*

	£ s d
Abraham Scales	7 6 1
Samuel Johnson	5 17 1
William Allen	3 4 6
Michael Wright	10 18 6
Henry Colby and others	3 7 7
William Colby	18 6
William Balls	1 6
" "	1 0 9
C & B Tilmouth[64]	11 17 1

The above bills having been allowed, ordered that the treasurer discharges them.

R.H. Reeve
secretary

A meeting of the committee appointed for superintending the building of a life boat to be stationed at Pakefield holden at the Crown Inn Lowestoft 3rd March 1841
Mr Geo Everitt in the chair

| Lieut, Carter R.N. | Mr G.S. Gowing |
| Mr Wm Everitt | Mr Edmd. Norton |

and R.H. Reeve secretary

Ordered that Mr G.S. Gowing's[65] bills of **[60]** £40 3 0½ for Pakefield life boat and £3 9 6 for Lowestoft life boat be paid.
Resolved that notices be sent to Messrs Wright, Bemment and Scarll to send in their tenders for building the Pakefield boat house and that they send them in by the 16th March 1841. The plan and specifications to be seen at Mr Norton's office – and the meeting was adjourned to 17th March next.

A meeting of the committee appointed for superintending the building of a life boat to be stationed at Pakefield holden at the Crown Inn Lowestoft 16th March 1841
Present Charles Steward Esq. in the chair

Mr George Everitt	Mr S. Johnson
Mr G.S. Gowing	Mr Edmund Norton
Lieut. Carter	R.H. Reeve

At this meeting tenders for building the Pakefield life boat house was received from Mr Wright of Pakefield at £134 – and from Mr Bemment of Lowestoft at £141 10. Resolved that Mr Wright's tender be accepted and that the boat house be completed within three months from this time, and that Mr Wright shall be allowed three weeks to cut out the board for the work and three more for the completion of the boat house, and it was agreed that Mr Wright should pay £1 for every week over time

---

64 Sailmakers at Lowestoft.
65 George Sead Gowing, ropemaker.

allowed for cutting out the board, and £5 for every week over the time allowed for the completion of the work.
Ordered that John Martin's bill of £3 13 – **[61]** for leather buckets &c. be paid.

A meeting of the committee appointed for superintending the building of a life boat to be stationed at Pakefield holden at the Queen's Head Inn Lowestoft 17th June 1841

Present Charles Steward Esq. in the chair

Lieut. Carter	Robert C. Fowler Esq.
Mr George Everitt	Mr Samuel Johnson
G.S. Gowing	Abraham Scales
Edmund Norton	R.H. Reeve

At this meeting it was resolved that the outside of the new boat house at Pakefield be done with shine varnish.[66]
Mr Bemment having certified that the boathouse is properly built the treasurer is dierected to pay to Michael Wright the amount of his contract £134 – and also 30/- in addition as agreed for lowering the site of the boat shed.
The meeting also approved of a lease of the site of the boathouse from Mr Leggett to the treasurer for 21 years determinable as therein mentioned.
Ordered that Mr Bemment be paid 30/- for his trouble in making specification and superintending the building of the boathouse &c.

[62] A meeting of the committee of the Suffolk Humane Society held at Mr Norton's office on 13th day of September 1841
Present Mr William Everitt in the chair

Lieut. Carter	Mr J.S. Lincoln
Mr S.S. Johnson	Mr R.H. Reeve

Mr Carter having reported that the Lowestoft life boat was in good condition and fit for service, resolved that the management of the sailing and the command &c. of the Lowestoft & Pakefield life boats be left to Lieut. Carter R.N.

The Anniversary Meeting of the Suffolk Humane Society holden at the Queen's Head Inn in Lowestoft on Tuesday the 14th of September 1841
Sir T.S. Gooch Bart in the chair.

At this meeting the treasurer's accounts were audited and allowed and balance of the general account found to be in the treasurer's hands £265 9 7 and found to be due to the treasurer on account of the Pakefield life boat £7 7.
An application was made in behalf of Samuel Butcher and Benjamin Butcher for saving two men, James Ward and William Pipe, whose boat had sunk, the mast and bow only remaining visible. Ordered that they be rewarded with £1 – .

---

[66] Presumably a gloss varnish.

Captain Manby attended the meeting and **[63]** presented his papers and plans for the diminishing of shipwreck on the most fatal coasts of Suffolk and Norfolk, &c. Ordered that the plans be referred to the committee of the Suffolk Humane Society – resolved that a vote of thanks be given Captain Manby for his attendance.

The treasurer called the attention of the meeting to the state of the Shipwreck Association, when it was resolved that a statement be prepared by the committee and forwarded to the president of the Suffolk Humane Society for distribution to the heads of the county with a view of getting up a county meeting

Resolved that the boatman's salary at Pakefield be 2/6d. a week instead of 3/6 which William Colby accepted.

The following persons were proposed and admitted Members.

Revd. F.C. Fowler[67]	Annual Subscription	£1 1s.
Mr. William Youngman[68]	Do.	10s. 6
Charles Pearse	Do.	£1 0 0

The following gentlemen are appointed the committee for the ensuing year: treasurer, secretary, George Everitt, J.S. Lincoln, Lieut. Carter, G.S. Gowing, Revd J. Rumpf,[69] Charles Steward, Robert C. Fowler, Revd F. Cunningham.

[64] A meeting of the committee of the Suffolk Humane Society held at the Crown Inn in Lowestoft on Wednesday the 13th October 1841.

Present Charles Steward Esq. in the chair

Robert C. Fowler Esq.	Mr J.S. Lincoln
Revd F. Cunningham	" G.S. Gowing
Revd J. Rump	" Edmund Norton
Lieut. Carter	" R.H. Reeve

Captain Manby attended and presented his model and plans for the diminishing of shipwreck on the most fatal coasts of Norfolk and Suffolk but no definite determination was agreed to.

A meeting of the committee of the Suffolk Humane Society held at the Queen's Head Inn in Lowestoft on Wednesday 20th October 1841

Present Charles Steward Esq. in the chair.

A representation was submitted to the meeting relative to the establishment of the Shipwreck Association which was approved & ordered to be printed.

---

67 The Revd Frederick Cooke Fowler, of New Hall, Gunton.

68 William Youngman, brewer and maltster, Lowestoft.

69 The Revd John Rumpf, curate of Pakefield from 1836 to 1856, during which time the Revd Francis Cunningham was rector and also vicar of Lowestoft; Rumpf was rector of Pakefield from 1856 to 1859. Canon B.P.W. Stather Hunt records in *Pakefield: The Church and Village* (Lowestoft, 1938), p.60, that it was said that if a lifeboat call came during the service he would close his book and follow his flock to the beach to render what help he could. 'On one occasion, during a violent storm from the N.E., there were seven vessels on the shore opposite Pakefield. By repeated ventures through an appalling sea the crews of six were brought safely ashore. Totally exhausted, the crew of the lifeboat declared themselves incapable of further exertion. Then the Rector jumped into the boat and called upon any who were accustomed to the sea to follow him. At which the lifeboatmen again launched the boat and brought the seventh crew safely to land.'

The committee made a report on the proposed plans of Captain Manby.

Lieut. Carter reported the want of two double purchase blocks for getting up the life boat, which were ordered.

Mr Carter also reported that the life line to the Pakefield life boat was made of bad materials – ordered that Mr Gowing be applied to on the subject.

[65]   A meeting of the committee of the Suffolk Humane Society held at Mr Norton's office on Monday the 27th day of January 1842

Present Revd F. Cunningham in the chair

| Mr George Everitt | Robt C. Fowler Esq. |
| J.S. Lincoln | Edmund Norton Esq. |

The treasurer is ordered to pay the crews of each of the life boats on duty yesterday £5. – John Mills, master [of] the brig *Thomas Oliver* of Sunderland, on her voyage to London with coals – we were running back from the southward for shelter – the wind blowing hard & the state of the weather such that we could not make the buoys – we struck upon the Newcome Sands about 1 o'clock p.m., the following is the crew that were saved:

John Mills master	John Crowson mate
Edward Priestley	Thomas Lister
Thomas Watson	Robert Shears
Lune Wilson	Sandy McDonald

Seven of the men were saved by the Lowestoft life boat and John Crowson the mate by the Pakefield boat –

three without shoes, & three without jackets & two without caps.

Ordered that the necessary clothing to enable the men to start for Yarmouth on their way home be provided by Mr G.S. Gowing & Mr Samuel Johnson & that the men be relieved with 2/6d. each to pay their necessary expences to get them to Yarmouth out of the funds [66] [of the] Shipwreck Society

Suffolk Chronicle 19 February 1842

Sir. A somewhat imperfect account of a most daring and dangerous, but happily successful exploit performed by the sailors of our town and neighbourhood having appeared in some of the papers, an eyewitness and member of the Suffolk Humane Society begs to offer you full particulars ... On January 26 about 1 p.m. a vessel was observed to be in great distress on the Inner Newcome, the wind at the time blowing a hurricane and the seas running mountains high. The Lowestoft life boat with a crew of 19 men commanded by Lieut. S.T. Carter R.N. assisted by Mr. H.B. Disney, Trinity Pilot, was most promptly launched and proceeded to her relief. By the greatest exertion and skill a communication by means of a lifeline was established with the distressed men, who had fled to the rigging for safety – the sea at the time making quite over the life boat and filling her with water. Mr. Disney was washed overboard, but providentially did not lose his hold of the safety line and was again drawn into the boat. Seven out of the eight men on board the wreck were hauled through the surf into the life boat, a distance of perhaps 20 yards. At this time the anchor of the life boat came home, and with great difficulty and danger the boat was sheered under the bowsprit of the wreck when the cable was cut and she then proceeded to the shore full of water, landing her own crew and the rescued men in safety. Lieut. Carter was carried in a very exhausted state to a house near by, where the usual means having been resorted to he was after a few hours so far restored as to be able to return to his residence in a chaise. Meantime the Pakefield life boat manned by a crew endowed with the same high courage and good seamanship which had characterised their neighbours, made further effort to save the poor fellow who was left on the wreck, and had the happiness of rescuing him from a watery grave ... The Lowestoft life boat sustained damage to a considerable amount.

A meeting of the committee of the Suffolk Humane Society holden at the Crown Inn in Lowestoft on Thursday the 3rd day of February 1842.

Present Sir Thos Gooch Bart in the chair.

Lieut. Carter recommended that the Lowestoft life boat part with her present anchor and get a heavier one – and a second anchor be got for the Pakefield life boat – that the padding of the Pakefield llife boat be brought down to the water line – thinks two or more plug holes are wanting in Pakefield life boat.

William Colby reported that Pakefield life boat had lost her anchor and part of her cable, 80 fathoms left, also some of the small lines – some of the plugs were injured – clinks of tanks started – and canvass on outer boxes torn.

Ordered that the Pakefield life boat be thoroughly overhauled and the losses and damages reinstated and repaired under the direction of the life boat committee.

[67] The Anniversary Meeting of the Suffolk Humane Society holden at the Queen's Head Inn in Lowestoft on Friday the 26th day of August 1842.

At this meeting Mr Bransby Francis proposed by Lieut. Carter as one of the medical assistants of this Society, and seconded by Charles Steward Esq. and unanimously elected.

An application was made on behalf of James Goldsmith and William Welham for rescuing James Howden from drowning – ordered that they be rewarded with 5/- each.

Another application was made on behalf of the said James Goldsmith for endeavouring to save a girl from a watery grave – Mr Francis stated that he attended the girl for three hours but that all the means he used were unavailing; ordered that James Goldsmith be rewarded with 5/-. Lieut. Carter attended and stated that in using the usual restoratives in the above case that a blanket and other articles were spoiled at the Suffolk Hotel,[70] being the inn to which the girl had been carried – ordered that the landlord be paid 5/-.

An application was made on behalf of James Haynes for saving a child from drowning that fell into Lowestoft Harbour – Haynes not being present ordered that the case should stand over to the next anniversary meeting. An application was made on behalf of Cook, Peek, & Smith under the following circumstances: Mr Hodgkin, surgeon of [68] Lowestoft, was bathing with his sons in the bathing machines, one

---

70 At the foot of London Road not far from the harbour.

of Hodgkin's sons plunged in and got out of his depth. Cook, the bathing man, who could not swim, went out with a rope up to his neck, and by means of the rope succeeded in saving the boy, Peek keeping hold of the rope & pulling Cook and the boy ashore & Smith having obtained the rope; ordered that Cook be rewarded with 7/6, Peek 5/-, Cook 2/6.[71]

Ordered that Mrs Cooper be remunerated with 7/6d. for the services rendered to Lieut. Carter when brought there on the 26th of January last in a most exhausted state from the life boat.

Ordered that Thomas Cunningham be paid £1 towards his bill incurred by men saved from David Ricard's[72] vessel and others.

Ordered that the Lowestoft life boat crew be paid £4 and Pakefield crew £3 10 for [blank]

The following persons were proposed and admitted members:

William Masterson Junr.	10 6
William Pratt	10 6
Lieut Carter [Joachim *written in pencil*]	10 6
John Mackett	10 6

[69] A committee meeting of the Suffolk Humane Society held at Mr Norton's office in Lowestoft 26th October 1842.

Present Revd F. Cunningham in the chair

Mr George Everitt	Mr J.S. Lincoln
William Cleveland	E. Norton
George Gowing	R.H. Reeve

Isaac Woods Sinclair attended and stated that he was master of the brig *Don* of Witby [*Whitby*] laden with coals to London, that his vessel struck about eleven o'clock on Tuesday evening on the Newcome Sands and the waves were breaking over her – at ½ past 11 o'clock we left our vessel and took [*to*] our long boat, and were picked up by a Yarmouth fishing boat, George Mullinder, master, and landed at Lowestoft about an hour and a half after we had the left the vessel. The captain stated that himself & crew five in number had lost all their clothes – the following are the names of the crew:

Isaac Woods Sinclair, master – John Stevenson, seaman – John Throcknaughton, mate – John Readman, cook – John Potts, seaman – Jonas Brigham Appre[*ntice*]. Four of the crew were without jackets, one without either jacket or waistcoat & one without a hat – Ordered that the above persons be furnished with the jackets &c. stated to be wanted & that they also be relieved with 10/- to enable them to get to Yarmouth and that their breakfast bill be paid – Mr Gowing will be good enough to furnish the men as above to be paid from [70] the Shipwrecked Fund.

Chairman

N.B. at next meeting Mr Scales' claim for rent of boat stores to be taken into consideration.

---

[71] Presumably this is an error and Smith received 2s. 6d.
[72] Could this be the *David Ricardo* from which the crew were rescued in 1836?

A meeting of the [Com. of the *added in pencil*] Suffolk Shipwreck Society [*sic*] held at the office of Mr Norton at Lowestoft on 30th January 1843.
Present George Everitt Esq. in the chair

Mr G.S. Gowing	Mr William Cleveland
W.R. Seago	R.H. Reeve

Thomas Turnbole, master of the brig *The True Briton* of Sunderland, 255 ton, laden with coals, struck on the Cross Sands[73] yesterday morning about 3 oclock and went down in deep water about 20 minutes afterwards. Myself and crew, eight in number, then took to our long boat and were picked up at day light yesterday morning by a brig. The following persons composed the crew:

Thos. Turnbole, captain	Maymes Henderson, seaman	
Chayter Kirton, mate	James Gentle	do.
Robert Watends, cook	Robert Ennis,	do.
Joseph Blalock, second mate	Allen Smurford, appr[*rentice*]	
William Rabbidge, apprentice		

Ordered that the captain and crew be relieved with 2/- each to proceed to Yarmouth.

Chairman

[71] A committee meeting of the Suffolk Humane Society held at the Queen's Head Inn Lowestoft on 17th May 1843.

Present John Francis Leathes Esq. in the chair

Mr. William Everitt	Edmund Norton
Lieut. Carter	R.H. Reeve

Lieut. Carter attended and stated that the Lowestoft life boat had been sent off to a vessel called the *Farnacres*, 150 tons, laden with coals from Sunderland to London, Thomas Raine, master, she had been on the Home[74] Corton Sands 24 Hours. The ship's company, seven in number, also two passengers together with a yawl's crew of 16 who had been assisting the vessel, were all safely brought ashore; there was a great deal of sea alongside the vessel, in consequence of which the life boat was compelled to slip her cable and leave her anchor behind, some trifling damage was also done to the boxes –
Ordered that the crew of the Lowestoft life boat be paid £5, also the further sum of ten shillings for horse money.[75]

A meeting of the [Comt of the *added above in pencil*] Suffolk Humane Society held at Mr. Norton's office 18th May 1843

---

73 A bank north-east of Yarmouth extending about eleven miles southward from the Newarp lightvessel.
74 A variation of Holm.
75 Presumably use of a horse to tow the lifeboat along the beach into a position from which she could sail to the wreck.

Present Lieut. Carter   Mr G.S. Gowing
    Mr W. Everitt   William Cleveland
      E. Norton   R.H. Reeve

Thomas Raine, master of the ship *Farnacre* [*no final s*] **[72]** 150 tons, laden with coals, from Sunderland to London, attended and stated that his vessel struck on the Corton Sands about ten o'clock [*blank space*] on Monday last, two of the Lowestoft yawls came to our assistance. Yesterday about ten o'clock ship began to take in water and about 11 o'clock in the forenoon of the same day we hoisted a signal for the Lowestoft life [*boat*], at that time there was heavy sea. The Lowestoft life boat came and took me with my crew, six in number, also two passengers and 16 of the yawls' crews safely ashore. The following are the names of the crew: –

Thomas Raine, captain   –   William Smith, seaman
Andrew Wilson, mate   John Watson, apprentice
William Burton, seaman   James Grocers, do.

Ordered that the above crew be relieved with 10/- to assist them to Yarmouth & that the same be paid from the Suffolk Shipwreck Fund.

R.H. Reeve
secretary

[73] The Anniversary Meeting of the Suffolk Humane Society held at the Queen's Head Inn in Lowestoft on Wednesday 23rd August 1843

     Charles Steward Esq. in the chair
The following persons were proposed and admitted members:

The Revd Robert Collyer[76]	1 1 0 per annum
The Revd Thomas William Irby[77]	1 1 0
Frederick Morse Esq.[78]	1 1 0

An application was made on behalf of Benjamin Taylor, Joseph Allerton, and Mark Waters for saving the lives of James Saunders, William Neeve and Robert Watson; ordered that 10/- be given amongst them.
The following gentlemen were appointed the committee for the year ensuing: treasurer, secretary, Lieut. Carter, Frederick Morse, George Sead Gowing, Lieut. Joachim, Charles Steward, Esq. Captain Fowler, Revd F. Rumpf, Abraham Scales, Revd F. Cunningham & George Everitt.
Ordered that the crew of the Lowestoft life boat be paid £4 5 and that of the Pakefield [*lifeboat*] £3 15.

[74] A committee meeting of the Suffolk Humane Society held at the office of Mr Norton on the 31st day of October 1843
     The Reverend Francis Cunningham in the chair.
     Present Mr G.S. Gowing, Frederick Morse & R.H. Reeve.
Mr G.S. Gowing stated that a vessel called the *Helena* of Sunderland, tonnage

---

76   Rector of Gisleham.
77   Rector of Rushmere, resident in Kessingland.
78   Brewer and maltster, of Bell Lane brewery, Lowestoft.

170 and laden with coals, master, William Meshom, was discovered on Saturday morning on the Barnard – the vessel was going fast to pieces when the Pakefield life boat went off and succeeded in saving two of the crew only, all the rest were drowned before the life boat reached the vessel – one of the men that was saved died soon after he was landed and the other, Peter Hull, was so exhausted that all his clothes were obliged to be cut off.

Ordered that Peter Hull be supplied with necessary clothing which Mr Gowing was requested and he undertook to see properly done, and the committee further ordered Peter Hull to have 7/6d. to assist him to defray the expenses of his journey home.

A committee meeting of the Suffolk Humane Society held at the Queen's Head Inn on the 8th day of November 1843.

Charles Steward Esq. in the chair.

Present – Lieut. Joachim,[79] John S. Lincoln, Edmund Norton, Revd J. Rumpf & R.H. Reeve.

Ordered that the crew of the Pakefield life [75] boat be paid £5 for their services in going off [to] the vessell *Helena* of Sunderland on the 28th day of October last.

William Colby having stated on being required to give an account of the stores that Mr Carter had sold him part of the canvass with which the boat was covered – Ordered that the secretary write to Mr Carter on the subject.

William Colby having reported that the Pakefield life boat was lying on the beach ordered that she be kept there for the present.

A committee meeting of the Suffolk Humane Society held at the Queen's Head Inn Lowestoft on 15 day of November 1843.

Charles Steward Esq. in the chair.

Lieut. Joachim, Revd J. Rumpf & R.H. Reeve

The secretary having stated that he had written to Mr Carter and received an answer it was taken into consideration and William Colby having produced an inventory of the boat stores he was reprimanded for having purchased the canvass.

William Colby reported that on the 28 of October last a pair of creepers[80] were lost in rendering assistance to the vessell the *Helena* of Sunderland.

[76] A meeting of the committee of the Suffolk Humane Society held on 20 November 1843

Charles Steward Esq. in the chair.

Mr Hodgin[81] attended and stated that Mrs Joseph Gurney through Miss Browne of Pakefield had presented four Blankets for the use of the Society and which were directed to be kept at Mr Abraham Scales.[82]

R.H. Reeve
Secretary

---

[79] Lieutenant Richard Joachim, RN, in charge of coastguard at Lowestoft.
[80] Grapnels.
[81] Joseph Hodgkin, surgeon, of London Road, Lowestoft.
[82] Shopkeeper of Pakefield; White (*History, Gazetteer and Directory*) gives his name as Abraham Searles, but it is consistently given in the minutes as Scales.

The Anniversary Meeting of the Suffolk Humane Society held at the Crown Inn on 5 September 1844

Sir Thomas S. Gooch Baronet in the chair.

At this meeting Sir T.S. Gooch proposed and Charles Steward Esquire seconded the nomination of Lord Rendlesham[83] as vice president in the room of the late Admiral Irby[84] and he was unanimously elected accordingly.

At this meeting the case of Samuel Turrell was taken into consideration, resolved that he be rewarded with 10s.

The case of Robert Francis for saving a boy in the Lowestoft Harbour having been taken into consideration ordered that he be rewarded with 5s.

The case of William Norman for saving the live [sic] of [blank] Heavers having been taken into consideration ordered that he be rewarded with [blank].

The case of William Hales and two others [77] for picking up a man belonging to the *Cammilla* be rewarded with 5s.

Ordered that a new cable be made for the Lowestoft life boat.

The treasurer's accounts were examined and allowed and the balance found to be in his hands was £261 17s. 9d.

The committee appointed for the year ensuing were the treasurer, secretary, Frederick Morse, George Sead Gowing, Lieut. Joachim, Charles Steward Esq., Robert Fowler Esq., Revd J. Rumpf, Revd F. Cunningham, Abraham Scales & James Wigg Hickling.[85]

A meeting of the Suffolk Humane Society held at the Crown Inn on 4 February 1846.[86]

Sir Thomas Gooch Baronet in the Chair

Charles Steward Esquire   Captain Carter[87]
Revd F.W. Cubitt   Lieut. Joachim
Edmund Norton   G.S. Gowing
                and R.H. Reeve

It having been reported to the meeting that William Garner & James Man and another had saved Robert Jarvis from being drowned on Sunday last and satisfactorily proved to the meeting the fishing boat in which he was in having upset, resolved that they be rewarded with ½ guinea each for their efficient services.

---

83 Lord Rendlesham, of Rendlesham Hall near Woodbridge, was a descendant of Peter Isaac Thellusson who was born in Paris and settled in London as a merchant, being naturalised by Act of Parliament in 1762.
84 Rear Admiral Frederick Paul Irby, RN (1779–1844), Companion of the Bath, Rear Admiral of the White. A member of a long-established Lincolnshire family, he was the second son of Frederick, second Baron Boston (1749–1825).
85 James Wigg Hickling, attorney and agent for the National Provincial Bank at Lowestoft.
86 This appears to be an error for 1845.
87 Carter had at last been promoted to captain, retired, after more than thirty years as a lieutenant on half pay.

[78] A meeting of the [blank] held at the Queen's Head Inn Lowestoft on 3rd June 1846.[88]
Charles Steward Esq. in the chair.
Revd F. Cunningham
Revd J. Rumpf   R.H. Reeve
Edmund Norton

Mr Rumpf having represented to this meeting that three men were wrecked on the 21st instant and had no clothes –
Ordered that they have 30s. to obtain them the necessary clothes and that Mr Rumpf be requested to see the money properly applied.

At a meeting of the Suffolk Humane Society held at the Crown Inn on 6 July 1845
    John Francis Leathes Esq. in the chair
                            Present
Charles Steward Esq.   William Everitt
Lieut. Joachim   Geo. S. Gowing
Edmund Norton   R.H. Reeve
At this meeting it having been requested that the Society's life boats might be allowed to have an exercise day at Yarmouth Regatta[89] resolved unanimously that the boats be allowed to be exercised on that day and that the usual allowances and expences for exercising be paid.
    Resolved that Lieut. Joachim be requested to take the command of the Pakefield life boat and Mr Disney that of the Lowestoft.

[79] A meeting of the committee of the Suffolk Humane Society held at the office of Mr Norton on 17 July 1846
Present – Lieut. Joachim, G.S. Gowing, Captain Carter, William Everett, Edmund Norton & R.H. Reeve
It having been reported to this meeting that it was likely there would be some dispute in selecting the crew of the Lowestoft life boat for the exercising day at Yarmouth Regatta.
Ordered – eight men be selected from the Old Company and eight from the New and that Thomas Ellis be steersman.[90]

The Annual Meeting of the Suffolk Humane Society held on the 28 day of August 1845
Charles Steward Esq. in the chair
                            Present
    Revd. E.M. Love   Robert C. Fowler Esq.
    Mr. G.S. Gowing   R.H. Reeve Esq.

---

88 This also appears to be an error for 1845.
89 The Yarmouth North Roads regatta. In 1845 a special event was held at this regatta for lifeboats from the Norfolk and Suffolk coast, the boats being put through various tests under sail and oar.
90 The Old Company and the New Company were the two beach companies then operating at Lowestoft.

William Cole Esq.[91]   Captain Carter
Edmund Norton Esq.
Lieut. Joachim

The following members were proposed by Captain Carter and seconded by Mr G.S. Gowing – and unanimously elected.

The Revd Robert Carter	Annual Subscription	1. 1. 0
James Hodges Esq.	do	10. 6
Mr Robert Johnson	do	10. 6
The Revd R. Young	Donation	10.

Proposed by Mr R.H. Reeve and seconded by Mr G.S. Gowing that Mr Prentice[92] be elected an honorary member of this Society as surgeon; he was elected accordingly –

[80] Mr Harmer proposed by Mr Norton and seconded that he be elected an honorary member in the place of Mr Primrose as medical officer – elected accordingly.

Proposed by Mr Gowing and seconded by Lieut. Joachim that Captain Jerningham[93] be elected an honorary member of this Society.

The case of John Francis the elder and John Francis the younger for saving the life of a girl named Cook, daughter of Benjamin Cook, bathing man, having been taken into consideration, Mr Barber who was present when the occurance took place not being at home – resolved that the son be rewarded with 2/6d.

The case of William Thompson and William Lewis for saving four men out of the jolly boat of a smack driven on the flat from the shore having been taken into consideration, resolved that they be rewarded with 30/- to be divided amongst the company.

The case of John Sparham for saving a child out of Mr Robert Tripp's pulk hole – depth 4 feet – rewarded with 2/6d.

The thanks of this meeting were unanimously given to Captain Carter R.N. for his efficient services and for promoting the objects of this Society.

Subscription lists ordered to be printed.

The following gentlemen were appointed a committee for the ensuing year: the treasurer, secretary, Robert Fowler Esq., Charles Steward Esq., Lieut. Joachim, F. Morse, Revd John Rumpf, G.S. Gowing, Abraham Scales, William Everitt and Revd Francis Cunningham.

[81]   Ordered that the Pakefield life boat be paid £3 10. and 12/- for getting boat up [blank] 5/- extra and Colby 5/- extra.

James Smith appointed boatkeeper of Lowestoft life boat at salary of 2/- per week.

William Allen to be discharged as boat keeper on account of his old age and incapacity.

Charles Steward chairman
R.H. Reeve secretary

---

91 Possibly the author of *A poetical sketch of the Norwich and Lowestoft navigation works, from their commencement at Lake Lothing to the opening of the harbour, August 10 1831* (Norwich, 1833).

92 John Prentice, surgeon, of Lowestoft.

93 Commander Arthur Jerningham succeeded Captain Samuel Fielding Harmer as Inspecting Officer of Coastguard at Yarmouth. He was one of the committee appointed to report on the lifeboat models submitted for the Northumberland prize in 1851, and was responsible for introducing the trials of lifeboats at the Yarmouth North Roads Regatta in 1845.

The annual meeting of the Suffolk Humane Society held at the Crown Inn in Lowestoft on 25th August 1846.

Sir Thomas S. Gooch in the chair

Present

Chas. Steward Esq.

Robert Fowler Esq.

William Cole Esq.

George Seppings Esq.

Edmund Norton Esq.

R.H. Reeve

Lieut. Joachim

Proposed by Mr Norton and seconded by Mr Steward that Mr George Seppings be admitted a member of this Society – and he was elected accordingly by subscription 10/6d. per annum.

The cases of Nathaniel Colby and seven others for saving three men off Lowestoft who were upset out of a punt[94] having been taken into consideration, ordered that the crew be rewarded with £1. 0. 0.

The case of George Tubby for saving Edward Lydamore having been taken into consideration **[82]** ordered that [he] be presented with a Medal of the Society and rewarded with 10/-.

The case of Samuel Capps for saving a man that fell out of his fishing boat having been taken into consideration, ordered that he be rewarded with 5/-.

[These three cases are repeated in almost identical terms lower on the same page]

Ordered that the several [bills] on the society be paid, they having been first audited and allowed.

Ordered that the crews of the Lowestoft and Pakefield life boats be paid as usual.

The committee for the year ensuing to consist of the treasurer and secretary, Robert Fowler, Charles Steward, Lieut. Joachim, Revd J. Rumpf, George Sead Gowing, Abraham Scales, George Seppings, Revd Francis Cunningham,

Ordered that the Pakefield and Lowestoft life boats' crews be paid as usual for exercising.

The case of Nathaniel Colby[95] and seven others for saving three men who were upset out of a punt off Lowestoft having been taken into consideration, ordered that the crew be rewarded with £1 0 0.[96]

The case of George Tubby for saving Edward Lydamore having been taken into consideration ordered that he be rewarded with 10/- and in addition thereto be presented with a Medal of the Society.

The case of Samuel Capps for saving a man that fell out of his fishing boat having been taken into consideration, ordered that he be rewarded with 5/-.

Ordered that the Pakefield and Lowestoft life boats' crews be paid as usual for exercising.

The treasurer's accounts having been examined **[83]** there was found to be a balance of £274 15 6½ in the treasurer's hands.

---

94 A longshore fishing boat operating from the beach.

95 Coxswain of the Pakefield lifeboat.

96 This and the following three items have already been entered above and are repeated in error.

A meeting of the Suffolk Humane Society held at the Crown Inn Lowestoft on 14th October 1846   Present

John Francis Leathes Esq. in the chair
Revd Francis Cubitt   Edmund Norton
R.H. Reeve   G.S Gowing

Mr. Gowing reported the crew of the *Aimwell* to be in a very destitute state having only the clothes they stood in and that their vessel was lost yesterday morning at 7 o'clock off Happisburgh Sands –

Ordered that they [*be*] paid £3. 0. 0 from the Shipwreck fund –

Annual Meeting of the Suffolk Humane Society held at the Queen's Head Inn on the 26 August 1847

John Francis Leathes Esq. in the chair.

E.S. Gooch Esq.   Chas. Steward Esq.
Revd T. Sheriffe   Robert C. Fowler
R.H. Reeve   George Seppings
Revd. R. Gooch   A.S. Bence Esq.

George Tubby's case for saving [*blank*] Butcher who fell into the Lock pit[97] having been taken into consideration, ordered that he be rewarded with seven shillings and six pence.

[84] The case of James Dawson and Edward Barsham who together with Charles Gooch were in a fishing boat at the end of the North Pier on the 7th March last about 12 o'clock at night when Gooch fell overboard, depth of water about 10 or 12 feet. Ordered that the case stand over until the next meeting.

Ordered that Captain Manby's letter be acknowledged with a vote of thanks from this meeting.

Ordered that the crew of the Lowestoft boat be paid £3 10 0 for exercising this day, 10/- for getting boat up and boatkeeper 5/-.

The boatkeeper having reported that the pad boxes were a little chafed, ordered that the boat be overhauled and see what repairs are wanting.

Ordered that the crew of the Pakefield boat be paid £3 10 0 for exercising this day, 12/- for getting boat up and boatkeeper 5/-.

At this meeting the following members were proposed.

Henry A.S. Bence Esq. by Mr. R.H. Reeve, seconded by E.J. Gooch Esq, subscription £1 1 0.

Captain Andrews[98] by Mr Norton seconded by Mr R.H. Reeve, subscription £1 1 0.

The treasurer's accounts having been examined there was found to be a balance of £310 14 4 in the treasurer's hands.

[85] A meeting of the committee of the Suffolk Humane Society held at the Queen's Head Inn Lowestoft on 8th December 1847

Present   J.F. Leathes & Chas. Steward Esqs.
                E. Norton & R.H. Reeve Esqs.

---

[97] The sea lock at the harbour entrance.

[98] Captain W.S. Andrews, harbour master of Lowestoft and later manager of the North of Europe Steam Navigation Company.

The case of James Dawson & Edward Barsham reported at the last Anniversary Meeting for saving the life of Charles Gooch & which was ordered to stand over until the next meeting having been considered at this meeting, it was ordered that 2.6d. be paid to the said James Dawson.

A committee meeting of the Suffolk Humane Society held at Messrs Norton & Reeve's office 17th January 1848.

Present   Mr G.S. Gowing
Lieut. Joachim   Mr James Cleveland
Frederick Morse   Mr R.H. Reeve

Ordered that the crew of the Pakefield life boat be paid £5 for saving the crew (seven in number) from a vessel called the *Glenham Castle* wrecked on the Holme Sand and £1 for getting the boat up and down.

[86] Annual meeting of the Suffolk Humane Society held at the Crown Inn in Lowestoft on the 25 August 1848

Sir Thomas Gooch Baronet in the chair.
Revd F.C. Cunningham, Charles Pearse,
Robert C. Fowler Esq., Lieut. Joachim R.N.
Edmund Norton Esq., R.H. Reeve

At this meeting the following members were proposed and unanimously elected.
Captain Gooch R.N. by Sir Thomas Gooch Bart, seconded by R.H. Reeve, subscription £1 1 per annum
Messrs Lucas & Son[99] proposed by Mr Norton & seconded by Mr R.H. Reeve £1 1.
The case of Benjamin Butcher for having saved Mrs Hall's child who had fallen into a pulk hole at Lowestoft having been taken into consideration, ordered that he be rewarded with 2/6d.
The case of John Warford for having saved a son of John Swan, harbour pilot, having been taken into consideration, ordered that he be rewarded with 5/-.
The case of Daniel West for having endeavoured to save a man that had fallen into the lock having been taken into consideration, ordered that he be rewarded with 10/-.
The case of John Capps and others for [87] saving five men from the schooner *Susan* of Plymouth having been taken into consideration, ordered that Mr Capps and crew be recommended to the Suffolk Shipwreck Association.[100]

---

99   Lucas Brothers, a large firm of civil engineering contractors who had a depot in Lambeth, London, and a works in Belvedere Road, Lowestoft. The principals resided in Lowestoft.

100   This appears to be a reference to the Suffolk Association for Saving the Lives of Shipwrecked Seamen, founded in 1824 and the second county association of its kind; known as the Suffolk Shipwreck Association for short. However a further entry on the next page might refer to what is elsewhere called 'the Shipwrick Fund', used by the Suffolk Humane Society to assist shipwrecked seamen.

The treasurer's accounts having been examined there was found to be a balance of £333 17 3 in his hands.

A meeting of the Suffolk Humane Society held at the Crown Inn in Lowestoft on the 24 day of January 1849 pursuant to a notice given for that purpose.
Edward S. Gooch Esq. M.P. in the chair

Lieut. Joachim R.N.	Revd F.W. Cubitt
Revd F. Cunningham	Revd Thos. Sheriffe
Robert C. Fowler Esq.	Charles Pearse Esq.
Revd D.G. Morris	Mr Fredk. Morse
Mr William Youngman	Mr Thomas Morse
Mr William Cleveland	Mr George Seppings
Mr George S. Gowing	Mr William Cole
Mr Wm. Jones Woods	Revd John Rumpf.
Edmund Norton Esq.	R.H. Reeve
Mr Johnson	

Lieut. Joachim having stated that efficient services had been rendered by the Lowestoft beachmen in saving lives for which they received no reward, proposed by Mr Norton and seconded by Mr Cunningham that a committee be formed to communicate with the different Societies on [88] the coast to take such measures for bringing the state of the Suffolk <Humane Society> Shipwreck Association before the Grand Jury and county gentlemen of the court on the next Assizes as may be deemed most adviseable with a view to the establishment of the said Society; that the committee consist of E.J. Gooch Esq. M.P., Charles Steward Esq., R.C. Fowler Esq., Lieut Joachim R.N., Mr. Norton, Mr G.S. Gowing; carried unanimously.

Proposed by Mr Gowing that during the next three months that in any extreme cases where any boat shall go off for the purpose of saving lives and not for salvage the crew of such yawl or boat saving any lives in such extreme cases shall be rewarded with any sum which the committee of the Suffolk Humane Society under the peculiar circumstances may deem proper not exceeding five pounds; carried unanimously.

Proposed by Mr Gowing and seconded by Mr Seppings that the sum of £5 be paid to the crew of the *Salem*[101] eleven in number and the *Friendship*[102] seventeen in number for saving the crew of the Hanoverian schooner on the 22nd and saving the crew of the *Hearts of Oak* on the 25th December and carried – such reward to be considered as part of any reward that may be received from the Suffolk [*blank*] Society and that this reward is not to be considered as a precedent for future rewards.

Chairman

[89]   A meeting of the committee of the Suffolk Humane Society held at the Crown Inn Lowestoft on Thursday 6th September 1849
Present   Charles Steward Esq. in the chair

Revd F. Cunningham	Revd J.F. Reeve
George Seppings Esq.	Mr William Cleveland

---

[101] A gig belonging to the Old Company of beachmen.
[102] A yawl belonging to the Young Company of beachmen.

Mr George S. Gowing	Thomas Preston Esq.[103]
Edmund Norton Esq.	R.H. Reeve Esq.
Lieut. Joachim	

At this meeting the members present inspected the Lowestoft life boat with the assistance of Messrs. Brandford, Allerton and Sparham, boatbuilders, when the air boxes on the starboard side were found to be leaky.

This committee are of opinion that considering the present state and age of the boat it be recommended to the annual meeting of the members to take into full consideration whether it be not the most adviseable course to abandon the present boat and take such measures as may be adviseable for enabling the Society to build a new boat in her room.

The secretary is requested to give the usual notices for holding the annual meeting on Tuesday the 20th instant.

Charles Steward chairman

[90] Annual meeting of the Suffolk Humane Society held at the Queen's Head Inn in Lowestoft on the 20th September 1849.
Edward S. Gooch M.P. in the chair

Thomas Preston Esq.	Captain Gooch R.N.
G.S. Gowing	Charles Pearse
R.H. Reeve	Lieut. Joachim R.N.
Captain Smyth	George Seppings
Charles Steward	Dr. Whewell
F. Morse	W.Cole

The state of the Lowestoft life boat having been taken into consideration, resolved that considering the great age and present state of the Lowestoft life boat it is adviseable to condemn her and sell the hull, retaining the masts, tackle &c. for a new boat and that the hull of a new life boat be built in the place of the old one.

Resolved that the new boat be built on the model of the Caister life boat[104] and of the same length under the superintendence of a committee and that such committee be composed of Charles Steward & Thomas Preston Esq. & Lieut. Joachim.

At this meeting the following members were admitted:

Revd Dr. Whewell	1 1 0
John Kerrich Esq.	1 1 0
H. M. Leathes Esq.	1 1 –
General Cock	1 1

---

103   Thomas Preston, gentleman, of High Street, Lowestoft.
104   The boat referred to was designed by William Teasdel on the lines of the larger of the two Yarmouth lifeboats built in 1833, and built by Thomas Branford, a Yarmouth shipbuilder of some repute, under the superintendence of Captain Jerningham and Captain Spencer Smyth, the piermaster at Gorleston. She served until 1865 and saved many lives.

[91] Resolved that the committee appointed for superintending the building of the hull of the new life boat should have full power to obtain all necessary plans, admeasurements &c. to enable them to have a boat built upon the above plan, to select timber, enter into contracts and do all other matters necessary to the proper completion of such boat.

Resolved that a committee be formed for raising a subscription in aid of the funds for building the new boat, namely Revd F. Cunningham, Captain Steward, E.S. Gooch Esq. M.P., Captain Gooch R.N. & R.H. Reeve.

The following subscriptions were then announced to the meeting.

	£	s	d
E.S. Gooch Esq. M.P.	20	0	0
Sir Thomas Gooch Bart.	5		

Ordered that out of the Shipwreck Fund six suits of clothes be provided for the use of men landed from stranded vessels, such clothes to be lent them for the time being only and to be kept at Coastguard watch house.

Proposed by Mr Gowing, seconded by Mr Gooch, that during the next six months in any extreme cases where any boat shall go off for the purpose of saving lives and not for salvage the crew of such yawl or boat saving any lives in such extreme cases shall be rewarded with any sum which the committee of the Suffolk Humane Society under the peculiar **[92]** circumstances may deem proper not exceeding £5.

The following gentlemen were then appointed the committee for ensuing year – Edmund Norton, Richard Henry Reeve, Robert C. Fowler Esq., Charles Steward Esq., Lieut. Joachim R.N., Frederic [*sic*] Morse, Revd John Rumpf, George S. Gowing, Abraham Scales, Revd F. Cunningham, William Cole.

Ordered that the crew of the Lowestoft life boat be paid £1 as she was not able to go out and that the crew of the Pakefield life boat be paid £3 10 for exercising the boat this day, 12/- for getting out the boat [*blank*] and boatkeeper 5/-.

A meeting of the committee of the Suffolk Humane Society held at the Crown Inn Lowestoft on Wednesday the Ninth October 1849 at one o'clock in afternoon for the purpose of taking into consideration the best mode of disposing of the Lowestoft life boat.

At this meeting it was resolved that Charles Steward Esq., Thomas Preston Esq. and Lieut. Joachim R.N. be authorized to dispose of the Lowestoft life boat as they may deem most advantageous – the committee for **[93]** building the new boat.

This meeting requested Mr Cole to be good enough to order six suits of clothes for the use of men landed from stranded vessels ordered at the annual meeting, which Mr Cole undertook to do.

Charles Steward chairman

A meeting of the committee of the Suffolk Humane Society held at the Crown Inn on the Seventh February 1850 at eleven in the forenoon.

Ordered that the crew of the Pakefield life boat be paid the sum of £5 for saving the crew of the bark called the *Endymion*, eleven in number, but this sum is given on the distinct understanding that the boat be not again used on any excuse whatever for salvage, it being represented to this meeting that after the crew of the bark were brought on shore the life boat was again launched for the purpose of salvage.

And that 12/- be paid to the crew of the Pakefield life boat for getting her up and down.

Chairman

[94] A meeting of the committee of the Suffolk Humane Society held at Queen's Head Inn on the 13th May 1850
Present Mr G.S. Gowing, Frederick Morse, Lieut. Joachim R.N.
Lieut. Joachim having reported that the tanks of the Pakefield life boat and the interior of the boat required painting and thoroughly examining,
Ordered that it be done and that Lieut. Joachim be requested to superintend the same and to give the necessary orders to the tradesmen to do the same.

Chairman

Annual meeting of the Suffolk Humane Society held at the Queen's Head Inn in Lowestoft on the 27 day of August 1850

Edward S. Gooch Esq. M.P. in the chair.
Charles Steward Esq., John Garden Esq., Revd Jo. Cunningham, Mr G.S. Gowing, Lieut. Joachim R.N., Mr Edmund Norton, Captain Gooch, R.N., R.H. Reeve, Mr W. Cole.

Ordered that the following rewards be paid:
James Nobbs, Robert Cooper, William Gallant, John Saunders, John Rose, William Cooper for their exertions in saving lives of crew of French vessel *Pere Jolet* 10/6d. each.
[95] William Henry Sutton and [*blank*] Bond, Coast Guardsmen, for assisting in saving men upset from boat *Vixen*.   10/-
Robert Hood and two others assisting in above case with their boat   15/-
James Swan, Joseph Butcher and John Golder for saving [*blank*]   £1 1 0
Benjamin Day for saving a man in the harbour   5/-
John Spurden and David Spurden saving a Frenchman in the harbour   5/-
Joseph Dawson and [*blank*] Arnold for saving a man bathing   5/-

Mr Edward S. Gooch M.P. was elected a vice president in the room of Mr J.F. Leathes deceased.
The name of the new Lowestoft life boat having been taken into consideration, resolved that she be named the *Victoria*.
Ordered that a separate account be kept of the new boat and the proceeds of the old boat carried to that account.

Chairman

[96] A meeting of the committee of the Suffolk Humane Society held at the Crown Inn Lowestoft on the 6th day of January 1851

Present   Thomas Preston
                 Richard Joachim
                 Charles Steward

Ordered that the balance of Mr Sparham's bill for building the Lowestoft new life boat, viz. £104 4 5 be paid by the treasurer.
Ordered that the old anchor belonging to the Lowestoft boat be exchanged for a heavier one, the old anchor to be given in part payment.

Annual meeting of the Suffolk Humane Society held at the Queen's Head Inn Lowestoft on the 26th August 1851.

Present   Edward S. Gooch Esq. M P in the chair
                 R.C. Fowler Esq.   Charles Steward Esq.
                 Lieut. Joachim R N   George Seppings Esq.
                 Edmund Norton Esq.   R.H. Reeve Esq.
                 William Cole Esq.
                 Revd R.A. Arnold

Ordered that the Lowestoft and Pakefield life boats crews be paid the usual sums for exercising –
[97] Joseph Fletcher's case for saving James Thompson and William Tovell having been taken into consideration, resolved that they be rewarded with 15/- each.[105]
The following gentlemen were proposed as annual subscribers

J.H. Gurney Esq.	2	2	–
Revd H.F. Fell	1	1	–
H.T. Birkett Esq	1	1	–
Captain Small	–	10	6
Mr Thomas Barber	–	10	–

The secretary having laid Mr Neaves' letter before the meeting, resolved that the receipt of the same be acknowledged.
The accounts having been audited and allowed there was found to be in the treasurer's hands   £191 1 4½
The following gentlemen were then appointed the committee for the ensuing year:
Edmund Norton, R.H. Reeve, Charles Steward, R. Joachim, F. Morse, Revd J. Rumpf, G.S. Gowing, Abraham Scales, Revd F. Cunningham, William Cole.

                          E.S. Gooch
                          Chairman

---

[105]   There is some confusion here: only one name is mentioned but 'they' are rewarded with 15s. each. The other man, and the nature of the rescue, is revealed by the newspaper report printed below. Joseph Fletcher is 'Posh' Fletcher: see note 42.

Norwich Mercury 30 August 1851

On Tuesday the annual meeting of the Suffolk Humane Society was held at Lowestoft. The day was far from being suited to the occasion. Before, however, the appearance of the rain, the two life boats on the station, namely the Lowestoft and Pakefield boats, left the inner harbour, and according to custom proceeded to test their sailing qualities by an excursion to sea. Taking twenty or thirty visitors on board (including several ladies) they proceeded across the sands, and then returned. Both boats are of first rate character, that belonging to Pakefield specially so. On the return to the harbour, about twelve o'clock, it was intended to display the mode of communicating with ship-wrecked vessels by means of the Manby and other apparatus, but the leaden and gathering sky, from which began to pour a flood of rain which proved unremitting, dispersing the company assembled on the pier, induced the committee to forego that part of the day's proceedings. At two o'clock the committee and subscribers met at the Queen's Head Inn to audit the accounts and decide on rewards. E.S. Gooch, Esq., M.P., son of the president, Sir Thomas Gooch, Bart., took the chair. The yearly account, as made up by the treasurer, Edward Norton, and the secretary, R.H. Reeve, Esqrs., stands as follows:

August 1850. – Balance in treasurer's hands	333 13 0	Cost of new hull and fittings, Lowestoft boat	210 10	
Subscriptions and arrears received	77 2 0	Expenditure Pakefield boat, including boatkeeper's salary and allowance to boat's crew	19 8	3½
Donation	1 0 0	The like expenditure for Lowestoft boat	14 13	6
Subscription towards new boat	26 1 0	General expenses	3 15	11
Bank interest	3 3 6	Reward to Joseph Fletcher and Wm. Smith, for saving Wm. Lovell in Lowestoft Roads	1 10	0
			249 19	1
		Balance in treasurer's hands	191 1	4½
	440 19 6		440 19	6

The reward of thirty shillings given to Fletcher and Smith, two boatmen, was for promptitude and energy displayed in saving the life of an old man who, with his companion, was capsized from his boat off the Barnet [*error for Barnard*] sand. Fletcher and Smith were fishing, but perceiving the accident they proceeded with all speed to the spot, and saved the elder of the two. The other man was rescued by a light schooner, but a sudden squall throwing the vessel suddenly into the wind, she shot past the second drowning man. This is not the first time Fletcher has saved human life by his own exertions, and for which he has previously received the gratuity of the Association – At the close of the busines the committee and friends dined together at the Queen's Head, E.S. Gooch, Esq., presiding.

We have much pleasure in stating that a Coast guardman of this station, named Cyprian Hatt, was a competitor for the prize offered by the Duke of Northumberland, for the best model of a life boat. The only instrument used in the construction of Hatt's boat was his clasp pocket knife; and although his boat did not reach the point of perfection – 100, it is gratifying to state that it so far secured the approbation of the committee as to be classed with the 29 others selected to be sent to the Crystal Palace for exhibition, where it may now be seen.

Committee meeting of the Suffolk Humane Society held at the Queen's Head Inn Lowestoft 8th January 1852.
Present Lieut. Joachim Chairman
    William Cole, R.H. Reeve, G.S. Gowing

Lieut. Joachim having reported that the cable [98] of the Pakefield life boat was much worn and had parted in two places and was not therefore in an efficient state for service, ordered that a new cable be furnished and the old one disposed of and that Mr Gowing be requested to furnish the same.

It being reported that the Pakefield life boat was on Saturday evening last taken off a second time for the purpose of salvage, ordered that they be not paid the usual sum for saving life and for going off.

R. Joachim
Chairman

Committee meeting of the Suffolk Humane Society held at the Crown Inn Lowestoft on the 26th day of January 1852.

Present   Charles Steward Esq. in the chair
    Thomas Preston   R. Joachim
    R.H. Reeve   Revd F. Cunningham
    Edmund Norton   Revd J. Rumpf
    G.S. Gowing   William Cole

It being reported that the Pakefield life boat had been used for purposes of salvage and this having been confirmed by Mr Rumpf and others and the men having admitted it, but that they did it as their own men had used their yawl to save life, the coxswains were desired and directed not to use the boat for salvage in future on any pretence whatever. £5 12 – [99] was awarded to the Pakefield crew for going off to save lives from a vessel which afterwards sunk.

Charles Steward
Chairman

Committee meeting of the Suffolk Humane Society held at the Queen's Head Inn Lowestoft on Wednesday 9th June 1852.
    William Cole Esq. in the chair
    Lieut. Joachim   R.H. Reeve

Lieut. Joachim having reported that the state of the tanks of both boats was such

58

that they required close examination and certain repairs and painting, ordered that such repairs as are necessary be done under the superintendence of Lieut. Joachim who is requested to superintend same.

Mr Joachim further reported that the repairs required to the Pakefield boat were occasioned by her having been used in the late salvage case in January last.

Chairman

[100] The annual meeting of the Suffolk Humane Society held at the Royal Hotel[106] Lowestoft on the 7th day of September 1852.

Present Sir Edwd. S. Gooch Bart. in the chair

Revd E.M. Love	R.C. Fowler
Thomas Preston	George Seppings
Lieut. Joachim	Robert Fiske
Revd R. Gooch	George S. Gowing
Edmund Norton	R.H. Reeve

Proposed by Revd E.M. Love, seconded by Mr G.S. Gowing, carried unanimously that Sir Edward Sherlock Gooch Bart. be president in the room of the late Sir Thomas S. Gooch deceased.

Proposed by Mr G.S. Gowing, seconded by Mr Seppings, and carried unanimously that Samuel M. Peto Esq M P.[107] be elected a vice president.

Proposed by Mr Norton, seconded by Mr Reeve and carried unanimously that F.H. Irby Esq. be elected a vice president.

The case of George Milligan and Joseph Grimmer for saving Ann Smith, a woman who had fallen into the lock on 16th August, having been taken into consideration, ordered that they be paid 2/6 each.

The case of William King and John Manship for saving William Butcher and son and James Clarke in October 1850 having been taken into consideration, ordered that they be paid 5/- each.

The crew, seven in number, of the fishing [101] boat called the *Prima Donna* of Kessingland for saving the crew of the barque *Latona* aground on the Barnard Sand having been taken into consideration, ordered that the crew of the *Prima Donna* have 50/- to be divided between them.

The case of George Ellis for having on 28th August last saved a boy named Anguish having been taken into consideration, ordered that he be rewarded with 5/-.

The case of the Pakefield yawl having been taken into consideration, ordered that unless a list of the crew of the yawl be given in, the case cannot be considered.

The case of John Swan for saving the life of William Carver having been taken into consideration, ordered that he be rewarded with £1.

---

106 Built in 1848–49 by Lucas Brothers for Samuel Howett as part of the suburb of south Lowestoft then being developed by Samuel Morton Peto.

107 Samuel Morton Peto, railway contractor, of Somerleyton Hall, later knighted in recognition of his work in providing a railway in the Crimea to carry supplies to the front.

The following gentlemen were appointed the committee for the ensuing year: Edmund Norton, R.H. Reeve, R.C. Fowler, C. Steward, Lieut. Joachim, Thomas Preston, F. Morse, Revd J. Rumpf, G.S. Gowing, Abraham Scales, Revd F. Cunningham, W. Cole.

E.S. Gooch

Chairman

[102] A committee meeting of the Suffolk Humane Society held at the Crown Inn Lowestoft on 29th September 1852 called relative to correspondence between Lieut. Joachim and Andrew Johnston Esq. relative to the loss of the *Weir Packet*.

Present Charles Steward Esq. in the chair

Revd E.M. Love	Revd J.F. Reeve
Robert C. Fowler	Thomas Preston
William Cole	Lieut. Joachim
G.S. Gowing	R.H. Reeve
Edmund Norton	Revd F.W. Cubitt

The correspondence between Mr Johnston and Lieut. Joachim was read by the secretary.

William Allington Peck, master of the *Second Adventure* of Lowestoft, states on the 12th instant I was coming in from sea, we passed the vessel *Weir Packet* about a warps length from the dumb buoy or South Holme, near enough to throw a biscuit on board – we came to the leeward of her – the wind east by south – did not observe any thing amiss nor see a squall – I read her name the *Weir P*acket – I saw a man come out of her and light his pipe – she went out of the Stanford Channel, this was between three and four – she was in smooth water under the lee of the Holm – it was blowing very hard – it was about an hour before dark – [103] the vessel was going free – with the wind on her beam – we came to the harbour and when we got within the piers – I heard the people on the pier call out that the vessel was running up a flag – my impression was that she shipped a heavy sea and sprung a leak – there was at that time a strong ebb.

Henry Tilbury, Coastguardsman, states – I saw a fishing boat pass the lost ship – I was at the Battery – I saw a flag in the rigging about four o'clock within a few minutes either way – it was about a quarter of an hour after the boat passed her – my attention was called to the vessel, saw her haul her mainsail up – I thought she had split her mainsail, she never struck at all, I was watching her – she continued her course about ½ an hour after I saw the flag and then she bore up for the sand and saw the sea break over her on the sand – a very heavy sea was running where she struck – I thought at the time the yawls could have saved the crew.

Thomas Ellis states – I was at the north pierhead – I saw the ship pass the Stanford Light in the proper channel about three o'clock I watched her till a heavy sea struck her, so heavy that I could hardly see her – I think it was about half an hour after she passed the Stanford Light – she felled her head yard and put her head about south – I saw two men one above another up the rigging and set a flag flying, one fastening it above and one below. She was about half a mile to the southward of the [104] black buoy; she continued her course from 10 minutes to a quarter of an hour after putting up the flag, standing to the southward – the ship then bore up to the broadest and shallowest part of the Newcome and struck there – she would strike in about three fathoms; there was broken water to the westward but none to the eastward as the sea broke after –

The yawls launched directly after the flag was hoisted – the yawls went to the leeward of the Newcome, there was too much sea to get out at the Stanford Gat – they no doubt expected the ship would have gone on and they would have got her off Pakefield – she was about 10 minutes from bearing up to striking

It was impossible for yawls to have got out of the Stanford Channel – they could not lead out.

If the life boat had been got out by tug or otherwise [I believe there *crossed out*] I believe there was so much water the men would have been washed out – had I felt there was any chance for making use of the boat I should have gone down to the life boat house.

Before dark I saw five men go up the starbord rigging. I was therefore satisfied she was full of water.

The life boat could not have gone out of the Stanford – the sea went over the fore yard of the ship – when the yawls went off it was not a steamboat case.

George Page of the lower light house[108] states: I saw her [105] strike when she went on the sand – sunset 10 minutes past four – I was lighting the lamps at the time she went on the sand – I was lighting somewhat earlier as it was a dark evening – I thought the yawls would have boarded her till she altered her course and struck.

Lieut. Joachim states: I went to the Beach a quarter after four; I had a key but was informed by the boatkeeper that the boathouse was unlocked – I know the sands well, I think it was a desperate case – there were few men to be had – the yawls' crews were out and numbers of men were out in the fishing – till the yawls' crews came in the boat could not have been got off.

John Smith: I am son of the boatkeeper – as soon as the yawls were off, I ran up for the keys – I gave Henry Spurden the keys to open the door of the boathouse and I ran to Lieut. Joachim to tell him, he was coming out of his door.

Charles Steward
chairman

[106]   A meeting of the committee of the Suffolk Humane Society held by adjournment at the Crown Inn Lowestoft on the 13th day of December 1852

Present   Charles Steward   chairman
         Thomas Preston   G.S. Gowing
         Lieut. Joachim   William Cole

---

108   The Low Light was on the Denes below the cliff and was designed to be moved along the beach when necessary. When kept in line with the High Light it brought ships through the Stanford Channel.

Daniel Cooper states: I am master of the fishing boat *Reliance*. On Friday the 12th November last I came in through Saint Nicholas Gat in company with two other vessels and the *Weir Packet*. I brought up under the Holme Sands, the night being very unfavorable, as did the two other vessels – I was surprised to see the *Weir Packet* stand on, on account of the state of the weather, the night promising to be coarse and bad. I remarked this to the mate, who quite agreed with me – when we brought up it wanted about quarter to four o'clock, at this time it was blowing a strong gale with wind east south east with a very heavy sea – the vessel took the Stanford Channel – I observed something amiss, she hove to the wind – I thought she struck on the sand head. This was about ten minutes afterwards, she proceeded on her passage for about five or ten minutes when she bore up and ran upon the Newcome Sands – I could not see whether she made any signal of distress – in 20 minutes after she struck it became so dark that I lost sight of her – as soon as she struck I knew it was impossible for any boat to save the crew and I knew the whole crew **[107]** must be lost – I am quite sure if people had been ready, on shore, to start, it would have been impossible to have got there before dark, and the steam boat would not have got the life boat there in time – it was about a mile from where I brought up and anchored to where the *Weir Packet* was on the sands.

It is my opinion from the *Weir Packet* being on the weather side of the sands on which a heavy sea was running that no life boat could have saved [*the*] lives of the crew as it would have been to the inevitable destruction of the life boat and the crew as the boat would have had to have gone through near a quarter of a mile of broken water, it being the broadest part of the sands – I saw the yawls go off, the ship was not aground then and did not get aground for some minutes after.

Robert Manthorpe states: I live at Kessingland. On Friday evening the 12th November last I went on to the beach about five minutes after six o'clock. I saw several loose boards wash ashore – about five or ten minutes afterwards I saw the stern frame, pump, and jib boom come on shore – about quarter before seven about eleven beams washed on shore – it was about three miles from where the vessel laid to the shore.

This meeting is of opinion, after having heard the evidence as to the time of the occurrence which was clearly 4 pm and not 3 pm as had been stated, having heard that Lieut. Joachim was in discharge of his voluntary duties a few minutes after 4 pm and having heard from men of undoubted bravery and experience, men who had never thought of making **[108]** a bargain before they went to save life, but who, on the late occasion could not be induced by liberal offers to hazard their lives, that it was impossible to have rendered assistance from the extremely dangerous position of the vessel, having heard all these things, is of opinion that the anonymous remarks which have appeared are most unjust and unfounded and were evidently written by a man in total ignorance of the facts of the case, and have no hesitation in saying that it would have been most imprudent to have sent off the life boat when it had become a life boat case, the danger attending such a step was more than imminent and it is the opinion of this meeting would have involved certain destruction to all on board.

Ordered that Robert Manthorpe be paid 5/- for his loss of time.

Charles Steward
chairman

A meeting of the committee of the Suffolk Humane Society held at the Crown Inn Lowestoft on the 10th day of January 1853

Present   Chas Steward Esq. – chairman
          Thomas Preston   Revd F. Cunningham
          G.S. Gowing   Lieut. Joachim
                    William Cole

Lieut. Joachim reports that when the Pakefield boat went off on Thursday last, she was much [109]damaged and that he had ordered her to be repaired immediately.

Ordered that the Pakefield and Lowestoft life boats be paid the usual sum for floating.

Ordered that four skids with rollers109 be provided for the Pakefield life boat and that Mr Joachim be requested to order them.

Moved by Mr Gowing, seconded by Mr Cole and carried unanimously that life belts be ordered on Captain Ward's principle.

Moved by Mr Gowing, seconded by Mr Reeve and carried unanimously that a register be kept of the times the life boats go off on actual service, and the name of the vessel she goes off to and the service rendered.

Ordered that a life buoy be provided for each boat.

Ordered that 15/- be paid for launching and hauling up boats.

                                        Charles Steward
                                        chairman

[110]   A meeting of the Suffolk Humane Society held at the Crown Inn in Lowestoft, pursuant to notice on Monday the 4th April 1853.

Sir E.S. Gooch Bart. in the chair
          Revd E.M. Love   C. Steward
          Revd Thomas Sheriffe   Lieut. Joachim
          Henry Hodges   William Cole
          G.S. Gowing   William Cleveland
          Thomas Preston   R.H. Reeve
          James Peto110   R.C. Fowler

At this meeting the adviseability of removing the Lowestoft boat into the harbour having been taken into consideration, resolved that she be not removed for the present.

It having been reported to the committee that the Pakefield life boat had been used for the purpose of salvage on a recent occasion the case was fully considered and the captain of the Pakefield life boat was cautioned, and desired not to use the Society's boat for salvage purposes on a future occasion.

                                        E.S. Gooch
                                        chairman

---

109   What the beachmen called *skeets*, used for launching and retrieving the lifeboat.
110   James Peto, brother of Samuel Morton Peto (see note 107), lived at 17 Esplanade, Lowestoft.

[111] The annual meeting of the Suffolk Humane Society held at the Royal Hotel in Lowestoft on the 20th day of October 1853

Present Sir Edward S. Gooch Bart. in the chair

Captain Gooch	R.H. Reeve
Lieut. Joachim	William Cole
George S. Gowing	Thomas Preston

At this meeting it was ordered that Thomas Liffen be rewarded with 5/- for saving the life of [blank] Flaxman and William Rose with 5/- for saving William Tripp at the same time.

Ordered that Charles Liffen be rewarded with 3/6 for saving James Rose.

Ordered that John Neave and three others, for saving Cornelius Ferrett, Henry Spurden and James Yallop and another, be rewarded with £2 0. 0.

Ordered that Robert Welham and Matthew Wiley for saving a woman who had fallen in the lock, be rewarded with 3/- each.

Ordered that Benjamin Taylor & William Gurney for saving a man who had fallen out of a smack be rewarded with 5/- each.

Ordered that William Cook, Richard Saunders and George Ayres for saving three men in the Basin be rewarded with 15/-.

Ordered that John Lyon be rewarded with 3/- for saving a woman who had fallen into a well.

The committee were reappointed.

Resolved that Mr Teed be appointed a member of the committee.

[112] The following gentlemen were unanimously elected as members of this Society:

S.M. Peto	3	3	0
George Edwards[111]	1	–	–
Henry Hodges	–	10	–
George Teed	1	1	–

E.S. Gooch
Chairman

Annual meeting of the Suffolk Humane Society held at the Crown Inn in Lowestoft on the 6th day of October 1854 at two o'clock in the afternoon

Present Sir E.S. Gooch Bart. MP in the chair

Lieut. Joachim RN	C. Steward
William Cole	R.H. Reeve
George Teed	Robert Fergusson[112]

Messsrs Joachim and Preston having suggested that two tanks be provided for the

---

[111] George Edwards was an engineer who had worked on the building of Lowestoft harbour and on other projects in the Lowestoft area. In 1890 he was instrumental in obtaining the winding up of the Suffolk Humane Society.

[112] Robert Fergusson, bank agent, of 16 Esplanade, Lowestoft.

Pakefield boat to give more buoyancy to the boat, ordered that such suggestion be carried out.

Ordered that Joseph Fletcher and six others for saving four men from the brig *Snipe* of Faversham, stranded opposite Mr Birketts, be rewarded with 7/- each.

Ordered that the crew of the Stamford[113] floating light be rewarded with 5/- each for saving the [113] life of one of the crew of the *Kentish Lass* who had fallen overboard.

Ordered that William Read of Dover, fisherman, be rewarded with 2/6 for saving the life of Richard Stannard who had fallen into the water near Lowestoft Bridge.

Ordered that Thomas Swan be rewarded with 5/- for saving the life of James Heson who had fallen into the harbour where there was 15 feet water.

Ordered that William Sutherland of Lowestoft be rewarded with 10/- for saving the life of John Bly.

Ordered that Joseph Norman and two others be rewarded with 2/6 each for saving the life of Catherine Barber, a vagrant, who was attempting to drown herself off the beach, and who was in a state of intoxication at the time.

Ordered that Robert Gurney and another be rewarded with 4/- each for saving the life of Simon Howard who was upset out of a punt in the North Roads.

Ordered that the usual payment of 5/- to the Pakefield & Lowestoft boatmen be made.

The Letter from the Office of the Committee of Privy Council for Trade Marine Department dated 13th September 1854 having been taken into consideration, it was decided that the committee would co-operate with the Lords of the Committee of Privy Council in carrying out the objects mentioned in their letter and would be glad to hear further information as to the objects proposed and also as to the terms upon which the committee are to have their boats placed under inspection.

The committee feel the boats should remain [114] as their own property and that the Society should retain the entire control of their funds and of all persons that may be required to carry out the necessities of this Society.

Mr Fergusson was proposed and unanimously elected a member of this Society
      Subscription £1 1 0.

E.S. Gooch
chairman

A meeting of the Suffolk Humane Society held at the Crown Hotel at 3 o'clock in the afternoon of the 4th of January 1855 to take into consideration the propriety of removing the Lowestoft life boat into the Lowestoft harbor and other special business pursuant to a circular sent to all the members of this Society by post on the 30th of December last.

Present Sir E.S. Gooch Bart. MP in the chair

Thomas Preston R.N.	Lieut. Joachim R.N.
R.C. Fowler	Charles Steward
George Teed	George Seppings
R.H. Reeve	George Edwards
W. Cole	

---

113 Error for Stanford.

At this meeting it was unanimously resolved that it was expedient that the life boat should be kept in the Lowestoft outer harbor if the required accommodation could be procured.

And it was the opinion of this meeting that [115] it would be desirable to rent rather than to build a life boat house and the following gentlemen were appointed a committee to draw up a statement to be forwarded to the directors of the Eastern Counties, Norfolk and Lowestoft Railway and Harbour Company asking their co-operation to carry out the above views of the Society: –

George Teed, Esq.   Capt. Joachim R.N.
George Edwards, Esq.   Lieut. Preston R.N.

E.S. Gooch
Chairman

To the directors of the Eastern Counties, Norfolk and Lowestoft Railway and Harbour Companies. –

Gentlemen,

The committee of management of the Lowestoft Life Boat Society beg to draw your kind attention to the following circumstances which induce them to ask at your hands some special accommodation for their life boat within your outer harbor where she may be preserved from injury and from whence she may be quickly launched on those sudden emergencies requiring her services, that thus the difficulty the committee has had of late years in launching the boat may be obviated, and that she may again be the happy means (under Providence) of saving many valuable lives.

The Lowestoft Life Boat Society has existed [116] since the year 1806 and from that time up to 1849 has been the means of saving no less than 300 lives! It has been supported by all ranks and parties and until recently by the hearty co-operation of the beachmen of the town, who never hesitated to leave their comfortable homes to risk the dangers of the storm to save the life of their fellow creatures.

The old life boat being worn out in the service was, in the year 1850, superseded by the present boat at a considerable cost to the Society, and it is believed to be one of the best life boats in the world. This boat is kept in an excellent boathouse belonging to the Society, on the Beach opposite the town, but to launch it from thence into the sea requires not only 20 stout hearts to form the crew, but the united exertions of some 70 or 80 willing hands to push her off. These were never wanting for the occasion though the remuneration which the Society could afford for the assistance was necessarily small when divided amongst so many; indeed the service and the risk may be almost said to have been voluntary. However, about the year 1849 considerable disaffection was found to exist on the subject and it was only with great persuasion sufficient assistance could be obtained to launch the boat; till at length on the 7th day of October last past when a Norwegian brig was upon the sands, with the crew in imminent peril, and when the beachmen were entreated by the committee to launch the boat, they peremptorily refused [117] to do so, declaring further that they never would do so again. It was only after great delay that these

66

men were ultimately saved by the Pakefield life boat. The exertion of the Life Boat Society being thus entirely paralyzed much fear is entertained that other accidents may occur beyond their power of help.

It is impossible to avoid here stating that the only reason alleged by the beachmen for their conduct is that, the steam tugs attached to your harbor have interfered with their employment as salvage men.[114] However this may be, and however the committee may regret that this view exists, the feeling is quite beyond their control and they find themselves forced to look round to ascertain if it be not possible so to locate their boat that it shall still be available to save the life of the drowning mariner without the necessity of getting 100 beachmen in the humour to "pull all together".

With this view a meeting of the committee was held on the 4th instant, Sir E. Gooch in the chair. The state of matters with the beachmen being laid before the meeting, it was resolved that it was expedient that the life boat should be kept in the Lowestoft outer harbour if the required accommodation could there be procured – that it should be placed upon an inclined plane or slipway, so that when not in use it should be safe above the level of the highest tides and protected by a house, and so that when required it could, with a small amount of manual assistance, or by means of a steam tug be expeditiously launched. A convenient site has been suggested to be, at the **[118]** [at the *repeated*] back or north side of the old north pier east of the cattle sheds,[115] where a space of about 90 feet in length by 18 in breadth would be required. The cost of an efficient structure for the purpose would be probably from £300 to £400. The enclosed tracing will explain the sort of building. It was further the opinion of the meeting that as there might be objections to the Society spending so large a sum (could they raise it) upon the premises of the Harbour Company it might not be unreasonable to suggest that if the Harbour Company would build the boat house the Society would be willing to pay annually a reasonable rent for the same.

As secretary of the Society I am requested therefore to make this statement to the Harbour Company and to ascertain their views upon the subject, it being of course clearly understood that the Society will keep the entire management control and repair of the life boat in their hands as heretofore, and although the situation already referred to is considered desirable the Society would be happy to confer with your engineer as to any other site or arrangements which may possibly better comport with his view of your interests, but take the liberty of respectfully requiring your early attention to this memorial as it is impossible to say, in the present state of affairs, how soon some awful sacrifice of life may occur which the proposed arrangement might have prevented if carried out.

I have the honour to be
Gentlemen
Your most obedient servant
(Signed) R.H. Reeve
Secy.

Lowestoft
6th January 1855

---

[114] This assertion was the cause of serious friction between the beachmen and the crews of the tugs, leading to at least one attack when stones from the beach were used to bombard the tugmen.

[115] Built by Samuel Morton Peto as part of his improvements to Lowestoft harbour and used to accommodate cattle imported from Schleswig-Holstein.

[119]

In reply to the above memorial the following answer was received on the 14th January instant.

Extract from the minutes of the Board of Directors of Thursday the 11th day of January 1855

"Read letter from Sir Thomas Gooch and other gentlemen interested in the Lowestoft life boat –

Resolved that permission be granted for the life boat to be stationed in the harbor provided it is no expense to the Company".

London January 17, 1855

Dear Sir

Lowestoft Life Boat House &c.

I have perused the letter & papers in this matter transmitted to you by Mr Leathes and Mr Reeve and examined the site in the harbour where it is proposed the boat house and slip should be placed and am of opinion that the situation proposed is not the most judicious and ere long probably will become valuable and necessary for the Company to have the power of appropriating for commercial purposes situated as it is in proximity to the old and new fish markets and cattle landing. A position at the opposite side of the harbour next the new south pier would I think, in every respect be preferable – with reference to the desirableness of finding accommodation for the life boat within the harbour I have no doubt whatever, and believe that indirectly it would operate to the advantage of the Company. The question of expenditure I find has been disposed of by the Board minute of the 11th instant **[120]** and with which it appears therefore I have nothing to do, otherwise I should have suggested that a sum of about £200 would have been amply sufficient to have provided all the real accommodation that was necessary presuming that the deepening in the outer harbour will be proceeded with. I beg herewith to return the papers &c.

D. Waddington Esq. MP	I remain Dear Sir
Chairman	Your obdt. servant
Eastern Counties Co.	Peter Bruff

A meeting of the committee held at the Crown Hotel 12th February 1855
Present – Charles Steward Esq. in the Chair

Captain Joachim R.N.	Lieut. Preston R.N.
George Edwards	George Teed
R.H. Reeve	Captain Fowler

Read letters from Eastern Counties Railway Company of 13th January last and from Mr Bruff, engineer to the Company, of the 17th January last. Unanimously resolved

That it is the opinion of this meeting that the Eastern Counties Company ought to build the boathouse, being paid by this Society a fair rental for the same, and the committee appointed at the last meeting are requested to again apply to the Eastern Counties Railway Company on the subject through Mr Till.[116]

The secretary was directed to call a general meeting of the Society for Monday next at the Crown Hotel to take into consideration certain [121] propositions relative to joining this Society with the National Life Boat Institution.

Ordered – That the crew of the Pakefield Life Boat be paid £6 10. 0. having launched the boat twice when they saved lives from the [blank] wrecked at Kessingland on the eighth instant.

Charles Steward
Chairman

To the directors of the Eastern Counties Norfolk and Lowestoft Railway and Harbor Company

----

Lowestoft Life Boat House

Gentlemen
      I have laid before our committee your answer of 13th January and they direct me again to call your attention to the subject, expressing a hope that you may yet modify your determination of withholding any further assistance as to accommodation for their life boat.

The humane and national character of the objects of the Society it was reasonably hoped would cause the proprietors of the harbour of Lowestoft to sympathise with them, more especially as the crews of the vessels heading to your port must be the first to benefit, and in extending the limits of your harbor of refuge (as regards life at least) from beyond the piers of the harbour even to the whole of the roadstead.

[122]   That your engineer Mr Bruff is of this opinion is clear from his letter of the 17th January last on this subject, in which he states with reference to the desirableness of finding accommodation for the life boat within the outer harbour "I have no doubt whatever and believe that indirectly it would operate to the advantage of the Company".

The Life Boat Society did not and do not now ask for the gift of the value of the boathouse as they propose to pay a rent for the same, which if amounting to the interest of the money scarcely amounts to putting the Company to the expence as deprecated in your letter of the 13th January last.

In again alluding to the estrangement of the feeling of the beachmen which has unfortunately placed the Society in their present predicament and suggested the present plan, the Society by no means side with them; on the contrary, is as unable

---

116   See Introduction, pp.xlviii–xlix.

to agree with them as to control them; but the facts being so might alone it was hoped have suggested the co-operation of the Company.

The committee have considered the alternative of constructing a boathouse themselves and meet the following difficulties. Their income consisting of annual subscriptions barely meeting the cost of management they are entirely unable to meet the outlay. – If, by possibility the required sum could be raised it is surely a question of doubtful prudence, expending it on another's freehold.

It being impossible to construct a house standing [123] independently of the pier, piling, and other works of the harbour, and as it could only therefore be constructed under the immediate inspection and to the satisfaction of your engineer or his agents even possibly as to the outward appearance of the structure it would for many reasons be better that the whole arrangement should be in his hands, he merely furnishing such accommodation as would be efficient for the boat.

That the frequent changes perhaps necessary in the arrangement of the harbor and its works and new views which changes of circumstances suggest might render the site now to be selected for a boathouse inconvenient to the Company or possibly useless to the Society at some future time. That the Company have at their disposal pile engines, creosote works with other implements and means which would render the construction of a house and necessary piling much less expensive to them than to the Life Boat Society – If the object sought was for the private benefit of individuals this Society would not venture again to trespass upon your valuable time; but their object, the continued means of saving some ten lives per annum must be their excuse.

I have the honor to be
Gentlemen
Your obedient servant

Lowestoft
   16 February 1855

Signed R.H. Reeve
Secretary

[124] A general meeting of the members of the Suffolk Humane Society held at Crown Hotel Lowestoft on the 18th day of February 1855 called to take into consideration certain propositions relative to joining this society with the National Life Boat Institution and on other special business pursuant to circular dated 13th February 1855 posted to every member of the Society.

Present   Charles Steward Esq. in the chair

Revd E.M. Love	Edward Leathes Esq.[117]
George Teed Esq.	George Seppings Esq.
R.C. Fowler Esq.	George Edwards Esq.
William Cole Esq.	R.H. Reeve Esq.

The following letter from Captain Ward was read to the meeting

---

117   Edward Leathes, of Normanstone House, Lowestoft.

Royal National Life Boat Institution
14 John Street Adelphi
12th February 1855

My dear Sir

In reply to your letter of the 17th ult. relative to a proposed connection between the Lowestoft Life Boat Association and this Institution; I beg to acquaint you that I have no doubt any such proposition on the part of your Society will be favorably entertained by the committee.

A copy of a circular letter which is about to be sent to the several branches of the Institution relative to the future management of its life boat establishments, will be forwarded to you in the course of a few days, together with other documents [125] of the Society, after the receipt of which your committee will probably forward, to the address of the secretary of the Institution, a statement of their wishes and views on the subject, defining the manner and extent of the connection which they propose, and stating how far they consider the rules already framed for the management of the life boats which are the "bone fide" property of the Institution will be also available and acceptable to them.

I am, my dear Sir,
very faithfully yours,
Capt. Joachim R.N..   (signed) J.R. Ward

Whereupon it was resolved, that the rules not being ready this meeting should be adjourned to the 5th day of March next at the Crown Hotel at one o'clock.

Charles Steward
Chairman

The adjourned general meeting held at the Crown Hotel Lowestoft on the 5th day of March 1855 pursuant to the last meeting

Present   Charles Steward Esq. in the chair
    George Seppings Esq.   Captain Joachim R.N.
    William Cole Esq.   George Teed Esq.

At this meeting were read circulars from Marine Department, Board of Trade, and Royal National Life Boat Institution and copy of Royal Life Boat Institution [126] rules and regulations, and the same being duly considered,

Resolved that this Society do unite with the Royal National Life Boat Institution and that the secretary be requested to write to the Institution to such effect.

At this meeting Lieut. <Joachim> Hockley R.N. was unanimously elected an honorary member \of this Society and member/ of the life boat committee.

Charles Steward
Chairman

The annual meeting of the Suffolk Humane Society held at the Crown Hotel Lowestoft on Monday the 10th day of September 1855 at 2 o'clock in the afternoon.

71

127

Life boat during the absence of Captn Joachim and
if necessary to make arrangements for keeping her
in the Inner Harbour during the ensuing Winter

The Case of William Cook and two others having
been taken into consideration Resolved that they
be rewarded with 7/6

Resolved that William Cook be rewarded with
2/6 for saving a Lad who had fallen into the
Water

The Case of Daniel West for saving the life
of a Woman who had walked into the lock pit
having been taken into consideration and it
appearing that he had used great exertions
Resolved that he be rewarded with £1. 0. 0

The Case of William Stork and Robert Wright for
saving the life of Samuel Butcher having been
taken into consideration Resolved that they be rewarded
with 10/-

The Case of John Golden, Thomas Porter & Charles
Swan and Joseph Butcher, for saving the life of Wm Ayers having been taken into
consideration Resolved that they be rewarded with
10/-

The Case of William King for saving the life of
John Welham having been taken into consideration
Resolved that he be rewarded with 2/6

10. *Page 127 from the Minute Book for 10 September 1855, courtesy of the Lowestoft Fishermen's and Seafarers' Benevolent Society.*

Present   Sir E.S. Gooch Bart. M.P. in the chair

Captain Gooch R.N.	George Edwards
Onley J. Onley Esq.	Lieut. Hockley R.N.
W. Cole	Revd Sheriffe
R.H. Reeve	Revd E.M. Love
Charles Pearse	

The secretary was requested to write the secretary of the Eastern Counties Railway Company as to providing the <life> boat house for the lifeboat and unless an immediate arrangement could be made an experiment should be tried for keeping her afloat in the outer harbour.

It was unanimously resolved that Lieut. Hockley be requested to take charge of the [127] life boat during the absence of Captain Joachim and if necessary to make arrangements for keeping her in the outer harbour during the ensuing winter.

The case of William Cook and two others having been taken into consideration, resolved that they be rewarded with 7/6d.

Resolved that William Cook be rewarded with 2/6d. for saving a lad who had fallen into the water.

The case of Daniel West for saving the life of a woman who had walked into the lock pit having been taken into consideration and it appearing that he had used great exertions, resolved that he be rewarded with £1. 0. 0.

The case of William Hook and Robert Wright for saving the life of Samuel Butcher having been taken into consideration, resolved that they be rewarded with 10/-

The case of John Golder, Thomas Porter Jr., Charles Swan and Joseph Butcher \for saving the life of William Ayers/ having been taken into consideration, resolved that they be rewarded with 10/-

The case of William King for saving the life of John Welham having been taken into consideration, resolved that he be rewarded with 2/6d.

The case of Robert Hook[118] for saving the life of [128] Daniel Mitchells having been taken into consideration, resolved that he be rewarded with 2/6d.

The cases of Henry Colby & Samuel Reeve were adjourned.

Lieut. Hockley, who went out in the Pakefield life boat, having stated that she was leaky, he was requested with Mr Sparham to examine her and Mr Sparham was further requested to send in a report and estimate of the probable expense of putting her in an efficient state.

The following gentlemen were appointed the committee for the ensuing year: Edmund Norton, R.H. Reeve, Charles Steward, Richard Joachim, Frederick Morse, Revd J. Rumpf, G.S. Gowing, Abraham Scales, William Cole, George Edwards and Lieut. Hockley.

The several bills were produced and allowed subject to Captain Joachim's examination and approval.

---

[118]   Hook had been appointed coxswain of the Lowestoft lifeboat in 1853.

11. Robert Hook, who became coxswain of the Victoria *at the age of twenty-five in 1853. From an old postcard in the editor's possession.*

Ordered that Colby be paid the usual allowance of 5/- as boatkeeper.

E.S. Gooch
Chairman

[129] The annual meeting of the Suffolk Humane Society held at the Crown Hotel in Lowestoft on Monday the 25th day of August 1856 at Two o'Clock in the afternoon.
Sir Edward J. Gooch Bart   In the chair

Charles Steward Esq.   Robert C. Fowler Esq.
George Edwards Esq.   George Seppings Esq.
William Cole Esq.   Captain Joachim R.N.
Captain F.W. Ellis   R.H. Reeve Esq.

The case of William Sparham for taking the crew out of the sloop *Sarah* was rejected on the ground that it did not come within the rules of the Society.

The case of Samuel Folkard for saving Mr Jackson whilst bathing, having been taken into consideration, resolved that he be rewarded with £1. 0. 0.

The case of Thomas Butcher for saving a boy named Rose having been taken into consideration, resolved that he be rewarded with 2/6d.

Resolved that the cases of Henry Culley for saving Samuel Folkard and William Peek for saving Thomas Durrant be referred to the committee to hear the evidence in support of them.

Resolved that the committee be requested to use their best endeavours to increase the subscriptions of the Society.
[130]
The following gentlemen were appointed the committee for the ensuing year:
Edmund Norton, R.H. Reeve, Charles Steward, Richard Joachim, Frederick Morse, Revd John Rumpf, George Sead Gowing, Abraham Scales, William Cole, George Edwards, George Seppings and the commander, for the time being, of the Coast Guard.

Chairman

A meeting of the committee of the Suffolk Humane Society held at the Court House in Lowestoft on the 1st day of September 1856

[*blank*]   In the chair

Resolved that the case of Henry Culley for saving the life of Samuel Folkard be rewarded with £1. 0. 0.

Resolved that William Peek for saving the life of Thomas Durrant be rewarded with £1. 0. 0.

Chairman

[131] A meeting of the committee of the Suffolk Humane Society held at the Court House in Lowestoft on the 22nd day of September 1856 at 12 o'clock at noon pursuant to notice.

Charles Steward Esq.   in the chair

Robert Cook Fowler	Edward Leathes
Richard Henry Reeve	George Edwards
George Seppings	William Cole

At this meeting it was resolved that the objects of and the services rendered by the Society be made public and Mr Edwards undertook to draw a statement of the same on being furnished with the necessary information.

Charles Steward chairman

A meeting of the committee of the Suffolk Humane Society held at the Court House in Lowestoft on the 10th day of December 1856

Charles Steward Esq.   in the chair

Robert Cook Fowler Esq.	R.H. Reeve Esq.
George Seppings Esq.	Capt. Joachim R.N.
William Cole Esq.	George Edwards Esq.

A statement of the number of lives saved by the Lowestoft and Pakefield life boats to the 1st of [132] December 1856 drawn up by George Edwards Esq. having been placed before the meeting,
Resolved that the same be entered on the minutes of the Society.

Ordered that the report on the present state of the Society as drawn up by George Edwards Esq. be printed and generally circulated and that the members of the committee be requested to use their best exertions to obtain subscriptions.

That it should be suggested at the annual meeting that the Medals struck for the Society should be in certain cases awarded.

And it was proposed and carried unanimously that the thanks of the meeting be given to Mr Edwards for his kind exertions in favor of the Society in drawing up the returns and report.

Charles Steward chairman

Account of lives saved by the Society's life boats up to 1st December 1856 abstracted from minute books, cash accounts &c. &c. by Mr George Edwards.

Date	Lives saved		Remarks
1815 January 12	3		
1821 October 22[119]	12		Crews of two vessels
" December	2		
1823 November 1	10		Manby's apparatus used – Harmer *Venus*
forwd	27		

[133]

1823 November 1	--		Crew of *Isabella & Margaret* saved by Manby's apparatus
1825 January 1	2		
1828 April 3	--		4 crew of *Lucks all* saved by yawls – Carter says life boat
" May 17	10		Crew of yawl picked up by life boat
" December 3	6		From Captain Carter's list
1829 November 23	10		
" " 24	9		
1830 November 29	10		From cash account only
1832 November 21	11		
1833 February 21	5		Only 4 in minute book – Captain Carter's list 5
1835 Janry 19	2		From cash account & Carter only, *Bishop Blaize*
1836 February 6	8		*David Ricardo* 7 *Speedwell* 1
1837 January 14	12		Carter's list only – cash account says a fishing crew of *Prince of Brazil*
November 2	14		Crew of *Bywell* 9, passengers 4
1838 January 10	2	Pakefield Boat	
1841 October 20	8		
1842 January 26	7	1	
1843 May 17	9		
	& 16?		Crew of yawl had been to assist
" October 31		2	
1846 January	7		
1848 January 17	7		
1850 February 7		11	*Endymion*
1853 January 6		4	No particulars in books from Captain Joachim
1854 October 7		8	Brig *Dronningen* of Norway "
1855 February 12		11	"
" " "		6	"
" November 3	3		"
	185	43	
	43		
	228		

---

[119] These rescues, from the sloop *Sarah and Caroline* and the brig *George*, actually took place on 22 October 1820. See pp.xli–xlii.

[134]   A special general meeting of the Suffolk Humane Society held at the Court House Lowestoft for the purpose of electing a president in the room of the late Sir Edward Sherlock Gooch Bart.

Charles Steward Esq. in the chair

The secretary read the letter convening the meeting.

Mr Steward having read a letter from Lord Stradbroke, it was proposed by Mr Robert C. Fowler, seconded by Lieut. Preston and unanimously resolved that the Earl of Stradbroke[120] be elected president of this Society and the secretary was directed to write and inform his Lordship thereof.

Charles Steward chairman

A meeting of the committee of the Suffolk Humane Society held at the Court House Lowestoft on Wednesday the 7th day of January 1857 at 12 0'Clock at Noon.

Charles Steward Esq. in the chair

Robert C. Fowler Esq.   Capt Joachim R.N.

At this meeting Captain Joachim laid the following **[135]** report before the committee.

5th January 1857

Tennant of Stockton

Soon after 3 pm on the 5th January 1857 the wind then N.E. with a most severe snow storm, a brig on the Newcome with the sea breaking over her made signals of distress by placing a flag in her rigging when the Lowestoft life boat in charge of Captain Joachim was launched and went off to her and having let go her anchor to windward dropped down under her stern and had succeeded in taking on board half the brig's crew, when in a tremendous squall the life boat's cable parted, but having taken a strong rope from the brig as a guy was enabled to hold on till the remainder of the crew were taken in with the exception of the master, who in attempting to get into the boat was washed overboard and was with considerable difficulty recovered and hauled into the boat in a senseless condition, having more than once disappeared in the sea, after which the boat immediately made sail for the harbour for the more expeditiously landing the master who had scarcely shewn any signs of life, but by the prompt and unweried [*sic*] exertions of the surgeon (Mr Worthington)[121] on his landing succeeded ultimately in his resuscitation.

The qualities of the boat appeared to be every way satisfactory and the zeal and hardihood of the crew was most praiseworthy throughout the severe snow storm and much praise is due to **[136]** the people on the beach for their ready aid.

---

120   The Earl of Stradbroke, of Henham Hall, was Lord Lieutenant of Suffolk.
121   Mr William C. Worthington, surgeon, of High Street, Lowestoft.

The brig was named the *Tennant* of Stockton, 220 tons burden – James Low, master – with a cargo of timber from Dantzic to London – and a crew of eight hands.

Resolved that Captain Joachim be authorized to order a new cable for the Lowestoft life boat and any other necessary repairs to the boat – also to dispose of the old cable or any part of it which may be found unserviceable for the use of the boat.

Charles Steward   chairman

The annual meeting of the Suffolk Humane Society held at the Royal Hotel in Lowestoft on Wednesday the Twenty sixth day of August 1857 at Two o'Clock in the afternoon.

Charles Steward Esq. in the chair

R.C. Fowler Esq.	Revd Francis William Cubitt
Captain Joachim R.N.	Captain Hill M. Leathes
William Everitt	Charles Pearse
Henry Hodges	Edmund Norton

R.H. Reeve

At this meeting Mr Edwards produced the following list of new subscribers to the Society, that is to say,

[137] Name	Subscription £	s	d	Donation £	s	d
Allen George					5	
Barnard Wm. Vince					10	
Beane George		5				
Breame James		10				
Brewster Robert		10				
Burton Clement					2	6
Barber Thomas (Blundeston)	1	1				
Balls James		5				
Bradbeer Benjamin M.		10				
Chamberlin Robert	1	1				
Chambers William Henry					2	6
Cleveland George					10	
Clemence Jno. L.		5				
Coleman Mrs.		5				
Colman and Stacy		5				
Cooper Rev. William					5	
Coleman M.					2	6
Chater William					5	
Childers Mrs. Leonard				2		
Chaston Robert		5				
Crowe Thomas		10				

Davy Miss		10		
Dowling George		5		
Devereux Jas & Thomas		5		
Devereux John		5		
Eastaugh J.D. & G.		10		
Everitt William	1	1		
Everard Miss		10		
E.C.			2	6

[138]

Ferrett William			5	
Friend A			10	
Fisher James		5		
Gee I.			1	1
Glover George		10		
Gooding Miss		10		
Garden John	1	1		
Gowing J.J.W.	1	1		
George Johnson		10 6		
Henner Dr.			10	
Howett Samuel		1		
Harveys and Hudsons			1 1	
Hodges Henry		10		
Jones Dr. Henry Bence			3	3
Knights John		5		
Leathes Capt. H.M.	1	1		
Leggett			5	
Lett			1	
Livock William			2	6
Leathes Edward	1	1		
Meadows Daniel		5		
Maguire			10	
Moore Mrs.			10	
Morris Robert			5	
Miller Philip		10		
Nottidge Rev.			10	
Paull Frederick A.		10		
Parr	1	1		
Pratt Thomas		5		
Peto James Esq.	1	1		

[139]

Rawlinson Mrs.			10
Roddam Mrs.	1	1	
Rumpf Miss		10	
Smith Lady	1		
Shandler Miss			5

Simons Rev.		5
Simeons John	5	
Stebbings George W.	10	
Till Richard	1 1	
Tymms Samuel	10 6	
Thirtle James Farrer	5	
Webb Dr.		10
Watson Richard		5
Westaway Robert		10
Wingfield Col.	1	
Woodthorpe William	1 1	
Woodthorpe John	10 6	
Worthington William Collins	5	
Woods William Jones	10 6	
Woods Thomas Roe		5
Woods Henry Glasspoole	10	
Waddington H.S. Jr. Esq.	10 6	
Youngman William B.	5	

At this meeting it was moved, seconded and resolved that the seventh rule of this Society, namely "That every annual subscriber of ten shillings and six pence shall be a Governor and shall be entitled to vote on all the concerns of the Society" be rescinded – and the same was rescinded accordingly.

[140]   Mr Edwards then proposed that the following rule be substituted in lieu thereof, namely "That every annual subscriber of five shillings shall be a member of this Society and shall be entitled to vote at any general meeting of the Society" which proposition having been seconded was unanimously carried.

Moved, seconded and resolved that subscriptions of five shillings be gratefully received and that subscribers of that amount be members of the Society without being proposed and elected at the annual meeting as heretofore.

At this meeting the following rewards were given viz:

William Peek saving	Thomas Durrant	1. 0. 0.
Henry Culley	Samuel Folkard	1. 0. 0.
Richard Butcher	Clement Nottingham	5.
Samuel Mewse	Peter Smith	5.
Robert Swan	A boy in the harbour	5.
Jeremiah Gilby	William Dixon	2. 6
Harris Allerton	Vertue Saunders	5.
George Gay & 2 others	William Harvey	3.
John Jenkenson & 2 others	John Sherry	7. 6
Robert Welham & Folkard	Manby	5.

Captain Joachim R.N. informed the meeting that the £5 that used to be allowed to the crews of the boats for exercising was discontinued, the London Society having undertaken to pay the same.

[141] Mr Simeons of the Coast Guard was unanimously elected an honorary member of this Society

The following gentlemen were appointed the committee for the ensuing year viz: Edmund Norton, Richard Henry Reeve, Charles Steward, Robert Cook Fowler, Captain Joachim R.N., Frederick Morse, Revd John Rumpf, George Sead Gowing, Abraham Scales, William Cole, George Edwards, James Peto and John Simeons.[122]

At this meeting it was resolved that in future the annual meeting of this Society be held on some clear day and not on the "Licensing Day" as heretofore.

Resolved that Mr John Barber Jnr. be employed to collect the subscriptions and arrears and that he be allowed 2½ per cent therefore.

Charles Steward   chairman

A meeting of the committee of the Suffolk Humane Society held at the Court House on the 27th day of September 1857
Present   Robert Cook Fowler Esq. chairman
   Edmund Norton   Captain Joachim R.N.
         Richard Henry Reeve

Resolved that the sailmaker's bill amounting to £ [*blank*] for a new foresail to the Pakefield life boat previously ordered be forthwith paid.

Chairman

[142] A meeting of the committee of the Suffolk Humane Society held on the 17th day of March 1858
Present   Robert Cook Fowler Esq. in the chair
   Edmund Norton   Captain Joachim R.N.
         Richard Henry Reeve

Captain Joachim having represented that the Lowestoft and Pakefield life boats went off on the night of the eighth day of March instant when it was blowing a perfect hurricane for the purpose of saving life, it was resolved that the respective crews be recommended to the National Society, notwithstanding that they had received salvage for services rendered.

Chairman

On the request of Captain Joachim the following letter is entered in the minutes of this Society.

14 John Street Adelphi
London W.C. 1st April 1858
My dear Sir,
   I am directed to inform you that the committee have granted a reward of 15/-

---

[122]   Later referred to as John Symons.

each to the 19 men who put off in the Lowestoft life boat to the rescue of the brig *Oswy* of Shoreham £5 for launching the life boat and 5/- each to the crew of 5 men of the steam tug.

The committee have also voted 10/- each to the cew of 38 men of the Lowestoft and Pakefield life boats for putting off to the rescue of the *Orwell* and £10 **[143]** for launching the two boats.

Two cheques amounting to £49 10 are enclosed herein, the amount of which you will please to apply accordingly and to send me the men's receipt for their respective amount as per enclosed forms

Yours very truly,

Capt. Joachim R.N.   Richd. Lewis Secy.

A Meeting of the Committee of the Suffolk Humane Society held at the Court House in Lowestoft on the 7th day of July 1858

Charles Steward Esq. in the chair

Captain Joachim R.N.   George Edwards
Captain Leathes   Richard Henry Reeve
Edmund Norton   Thomas Preston

Captain Joachim R.N. read a letter from Mr Lewis dated 24th June 1858 relative to building a new boat house for the Lowestoft life boat, and the secretary was directed to make the following reply thereto.

Lowestoft 13th July 1858

Dear Sir,

I beg to inform you that your letter of the 24th June last has been laid before the committee of the Suffolk Humane Society together with the plan of a proposed new life boat house at Lowestoft and I am directed to state that the committee are willing to assent to the National Life Boat Association rebuilding the life boat house but they are of opinion that no rebuilding would be useful unless the opening of the **[144]** doors is made 10 feet wide – the committee also suggest that this would be best done by using moveable shutters after the fashion of shop shutters.

I am further directed to state that the committee will be disposed to render any assistance in their power as suggested by your letter in soliciting contributions from influential inhabitants of Suffolk, in any mode which may be adopted by your Society, but they at the same time desire me to state that from their experience in recent appeals to the county gentlemen they feel little confidence that aid to any useful extent can be expected from such application.

I remain, Dear Sir

Richard Lewis Esq.[123]   Yours very truly
Secretary   R.H. Reeve
Royal National Life Boat   Secretary
Institution

---

123 The second secretary of the Royal National Lifeboat Institution, Richard Lewis held that post from 1850 to 1883 and was largely responsible for putting the Institution on a sound footing. He was the author of the first history of the Institution, *History of the Life-Boat and its Work*, published in 1874.

A Meeting of the Committee of the Suffolk Humane Society held at the Court House in Lowestoft on the 4th day of August 1858

Present Frederick Morse Esq.   in the chair
       Capt. Joachim R.N.   Richard Henry Reeve

At this meeting sanction was given for adding an additional weight of iron to the keel of the Lowestoft life boat and also for her painting and overhauling and Captain Joachim was requested to give directions for the same, the London Royal National Life Boat Institution having intimated that they would bear the expense if necessary.

Charles Steward Chairman

[145] The annual meeting of the Suffolk Humane Society held at the Royal Hotel in Lowestoft on Thursday the Twenty sixth day of August 1858 at two o'clock in the afternoon.
Charles Steward Esq. in the chair.
Present
   Colonel Wingfield R.A.
   John Symons   Captain H.M. Leathes
   William Cole   Revd R.C. Denny
   Edmund Norton   Richard Henry Reeve

Read the minutes of the last annual meeting and of committees.

Captain Joachim reported that since the last annual meeting 18 lives had been saved by the Society, namely on the [blank] day of [blank] 13 lives from the *Oswy* which vessel was wrecked on the North Beach and on the [blank] day of [blank] 5 lives from the *Orwell*.

At this meeting the following rewards were given, viz:

Samuel B. Cook for saving [blank] Farrer who had fallen off the pier landing steps into the harbour	10/-
Nathaniel Barber for saving Charles Pipe	5/-
William Beaumont for saving Lydia Dade aged 14 years who fell off the landing steps into the harbour	10/-
James Chambers Mewse for saving George Smith	5/-
George Neaves for saving John Riches who had fallen over board from vessel into harbour	3/6

[146]   Ordered that in future the memorandum at the end of Notice of annual meeting as to paying subscriptions to Mr Norton at meeting be left out.

At this meeting the accounts of the Society for the past year were audited and allowed, the balance in hand amounting to £173. 7. 10.

A donation of £20 was voted by this meeting to the Shipwrecked Fishermen and Mariners' Royal Benevolent Institution from the funds of the Lowestoft Ship-wrecked Mariners Society.[124]

The following gentlemen were appointed a committee for the ensuing year, viz:

The treasurer, the secretary, Charles Steward Esq., R.C. Fowler Esq., Captain Joachim R.N., Frederick Morse Esq, Revd John Rump, [sic] Mr George Sead Gowing, Mr Abraham Scales, Mr William Cole, George Edwards Esq., James Peto Esq., Mr John Symons and Colonel Wingfield R.A. – three to be a quorum.

Captain Joachim proposed that the secretary be instructed to inform the committee of the Royal National Life Boat Institution that this Society is willing to contribute have [half?] the expense of building the new life boat house provided that such proportion does not exceed £100.

Moved as an amendment by Mr Edwards, seconded by Mr G.S. Gowing that this Society present the Royal National Life Boat Institution with a donation of 50 \Guineas/ and the same being put to the meeting was carried.

Ordered that Mr Robert Brewster,[125] Mr William Sparham[126] **[147]** & Mr Thomas Swatman[127] be invited to send in tenders for the erection of the intended new life boat house.

Charles Steward   Chairman

A meeting of the committee of the Suffolk Humane Society held the 15th day of September 1858

Charles Steward Esq. in the chair
R.C. Fowler Esq.   W. Cole Esq.
Colonel Wingfield R.A.   Mr Symons

Read letters from secretary Royal National Life Boat Institution of 7th & 10th September instant and the same having been discussed, resolved
[The rest of the page has been left blank]

[148]   A meeting of the committee of the Suffolk Humane Society held on the 22nd day of September 1858
Charles Steward Esq. in the chair
R.C. Fowler Esq.   Capt. Joachim RN
Revd John Rumpf   Mr Symons

The following letter to the secretary of the Royal National Life Boat Institution as to the new life boat house was read and approved and directed to be forwarded.

---

124   Could this be a reference to the Shipwreck Fund mentioned elsewhere?
125   Cabinet maker, High Street.
126   Joiner and builder, Duke's Head Street.
127   Bricklayer and builder, London Road.

Lowestoft 23rd Sept. 1858

Dear Sir,

I placed your letter of the 10th instant before the committee who were unanimous in their regret that any thing should have taken place here to cause an uneasy feeling either in Mr Cooke or yourself, particularly after the generous support that we have at all times received from your Institution – but had we not made known to you the remarks of the three builders who were requested to tender on the subject of the hipped roof your Institution might have deemed it an omission of our duty.

The committee are quite willing to give way as to the gable but the Beachman [*sic*] and all who work with them are still desirous of having the full breadth of house at the entrance, namely 20 feet – the committee must therefore beg that this point will be conceded.

The old boat house has been removed and the foundation of the new laid.

I herewith enclose the agreement signed as you requested.

As the contractor has taken the work **[149]** at a low sum, and as none of the committee appointed to superintend the works are builders I beg to suggest on their behalf that an architect should be employed to superintend the work and would \recommend/ Mr George Glover, a subscriber to the Society and who would under-take it at a trifling sum.

> I am Dear Sir
> Yours very truly
> (signed)   R.H. Reeve
> Secretary

P.S. as the building is commenced an early answer will oblige as to superin-tendant of the work.

To Richard Lewis Esq.

The following Gentleman [*sic*] were appointed a committee to superintend building of boat house: George Edwards Esq., Thomas Preston Esq., Frederick Morse Esq., Capt. Joachim RN, and R.H. Reeve Esq.

Charles Steward Chairman

A meeting of the committee of the Suffolk Humane Society held on the 22nd day of December 1858
Present Charles Steward Esq. in the chair
R.H. Reeve Esq & Captain Joachim

Resolved that the treasurer be authorized to pay the bill of B. Tilmouth amounting to £7. 5. 11. for a new mizen for the Pakefield life boat.

Charles Steward chairman

[151] At a meeting of the committee of the Suffolk Humane Society held on the 12th day of January 1859
Present Charles Steward Esq. in the chair
Richard Henry Reeve Esq.   Captain Joachim RN

86

At this meeting sanction was given for ordering a new bumpkin and irons for the same for the Lowestoft life boat.

Charles Steward   chairman

At a meeting of the committee of the Suffolk Humane Society held on the 25th day of May 1859

Present Charles Steward Esq. in the chair

James Peto Esq.	William Cole Esq.
R.H. Reeve Esq.	Captain Joachim RN

At this meeting sanction was given to have the two life boats overhauled and the necessary repairs performed to fit them for the ensuing season and Captain Joachim was requested to give directions forthwith for the same.

Charles Steward   chairman

[*At the back of the minute book a single page (150) bears a list of donations and annual subscriptions entered upside down. This is undated, but from the names entered it would seem to be from the early days of the Suffolk Humane Society and it would appear that it had been intended to enter subscription income each year at the back of the book; perhaps after this entry had been made it was decided to obtain a separate book for this purpose.*]

Donations	£	s.	d.
R. Kerrison Esq. Bungay	1	1	
T. Farr Esq. Cove	1	1	
F. Sandys Esq. Architect	2	2	
T. Rede Esq. Beccles	1	1	
Annual Subscriptions			
Lord Rous President	4	4	
R. Sparrow Esq. V.P.	3	3	
Revd. B. Bence T	2	2	
Revd. R. Lockwood T	2	2	
Revd. G. Spurgeon Lowestoft	2	2	
Capt. Hinton Lowestoft	2	2	
Davy Davy Esq. Yoxford	1	1	
R. Reeve Junr. Esq. Lowestoft	1	1	
Charles Pearce Gent. Carlton	1	11	6
Revd. Mr. Orgill Brampton		10	6
Mr. A. Payne Lowestoft		10	6
– T. Cunningham		10	6
– J. Elph Lowestoft	1	1	
– J. Davie		10	6
– W. Cooper Kessingland		10	6

87

–	H. Davey Surgeon Beccles	1	1
–	W. H. Crowfoot do do	1	1
–	T. Crowfoot Kessingland	10	6
–	T.S. Crowfoot do	10	6
–	E. White do	1	1
–	W. Crowfoot Surgeon Beccles	1	1
H. Bence Esq. do		1	1
Revd. W.T. Spurdens S do		10	6
Revd. M. Maurice S Normanstone		2	2
Revd. Mr. Ellison		1	1
Mr. Pyman Beccles		10	6
C. Arnold Esq. Lowestoft		2	2
Mr. Aldred Lowestoft		10	6
Revd. Mr. Nicholls Beccles		2	2
Isaac Blowers Esq.		1	1
Revd. Mr. Wood		10	6

THE MINUTE BOOK OF
THE SUFFOLK HUMANE SOCIETY

1890

[On the cover]

SUFFOLK HUMANE
SOCIETY
JULY 1890

[1] *Title Page*

R.H. Reeve Esq
was Treasurer to
The Suffolk Humane Society
when he died Oct 18. 1888
On application to his Executors
for the Books & Papers relating
to the Society, some of them
only were to be found, the third
Vol. of the minute Books (supposed
to be the last) being still missing
has necessitated the providing
The Present Volume
1890

[2]   Early in the year 1890 it occurred to Mr. Edwards[1] (an old subscriber & friend of the Suffolk Humane Society,) to ascertain the position of certain funds known to belong to it, and with this object applied to Mr C.T. Turner[2] one of the exors. of the late treasurer Mr Reeve –

The result as to the funds was satisfactory, but some of the books being missing, Mr E. was unable to discover who was the president or chairman or secretary or other more competent person to call together the surviving members of the Society – after consulting some of them he therefore determined to call a meeting himself as per the circular annexed –

CARLTON COLVILLE,

June 23rd, 1890.

SUFFOLK HUMANE SOCIETY.

DEAR SIR,   1873

You will recollect that in 1842 [*crossed out and 1873 substituted above*] we turned over the life boat part of our business to the Royal National Society, but retained in the hands of our then treasurer, the late R.H. Reeve, several hundred pounds, partly invested and partly in cash amounting probably at the present time to £700, laying uselessly at the banks for more than 15 years.

I find your name in the last available list of subscribers, and as I think you will agree with me that it is not desirable the above state of affairs should continue, I venture to ask you to meet other subscribers and myself at the Police Court, on Wednesday, the 2nd day of July, at 3 o'clock, to discuss the matter.

Yours very truly,

GEO. EDWARDS

[3]
Circular sent to subscribers
of one guinea & upwards

J.J. Colman	
R.C. Denny	concurs, prevented by sickness
W.H. Andrews	attended
William Woodthorpe	
C.J. Steward	
James Peto	attended
Thomas K. Woods	
William Birkbeck	concurs, previous engagement
Henry Birkbeck	
Gurneys & Co.	

---

[1]   George Edwards, JP, an engineer living at Carlton Colville.
[2]   Charles Thomas Turner, of 131 London Road, Lowestoft.

92

J.H. Reeve	concurs, prevented by sickness
Fred Morse	attended
S. Hewett	Decd. ?
H.M. Leathes	attended
George Edwards	Over –

[4] Subscribers of 10/-

Miss Davey	
E.H. Edwards	attended
W.J. Woods	Decd. ?
William Youngman	concurs, previous engagement
H.G. Woods	concurs, do do

Subscribers of 5/-

Arthur Stebbings	
T.E. Thirtle	attended
J.L. Clemence	attended

[5] In consequence of the Police Court being under repair Mr Ellen[3] was kind enough to allow the following meeting to be held at his office in Gordon Street

on Wednesday July 2[d] – 3 PM

1890

Present   Revd W.H. Andrews[4] in the chair

James Peto[5]	Lt.Col. Leathes[6]
George Edwards	J.L. Clemence[7]
Fred Morse[8]	E.H. Edwards

and T.E. Thirtle[9]

The chairman having stated his reminiscences of the separation of the Society from the Royal National Institution called upon Mr Edwards to make his statement –

Mr Edwards aplogised to the meeting for having taken the liberty of calling them together but hoped his explanation would be sufficient to justify his having done so. –

[6] Mr Edwards researched in detail the circumstances which led to the separation of the two Societies, and as far as he could state without reference to the minute books (which could not be found) the arrangement then proposed for future action of the Suffolk Humane Society.

As there were certainly funds remaining belonging to the Society. Mr Edwards

---

3   Frank Stratton Ellen, solicitor and clerk to the county magistrates.
4   The Revd William Hale Andrews, rector of Carlton Colville, 3 Kirkley Cliff, Lowestoft.
5   James Peto, JP, 17 Esplanade.
6   Lieut.-Col. Hill Mussenden Leathes.
7   John Louth Clemence FRIBA, architect and surveyor, 14 Marine Parade.
8   Partner in Morse and Woods' brewery in Crown Street.
9   Thomas Elven Thirtle, ironmonger, of 45 High Street, Whapload Road and Old Market Plain.

applied to Mr Turner, one of the exors. of the late Mr R.H. Reeve, for information as to their position and to inspect the books to discover if possible who would be the proper person to call a meeting of the subscribers to appoint another Treasurer in the place of Mr Reeve who held that office, unfortunately the more recent "minute books" were not to be found, so that it was not possible to discover either president, chairman or other person qualified to call together the surviving subscribers – this induced Mr Edwards to take that duty upon himself –

Mr Turner was good enough to give some information respecting the funds remaining to [7] the Society which confirmed Mr E's previous knowledge – These arose from three sources:

1st   in order of date – being a fund set apart by the Society in 1820 called "The Shipwrecked and maimed" This is in the Lowestoft savings bank, and amounts now with principal + interest to £290 13s. 2d. as so stated by Mr Allerton –

2d   Miss Crow left a legacy to the Society in 1856 of £200 and is now invested in 2¾ consols £214 14s. 8d. standing in the names of R.H. Reeve, Hill Mussenden Leathes and George Fowler.

3d   Cash in Gurneys Bank to the credit of the Society of £200 or more, this includes the sum of £116 5s. 3d. surplus balance of the Nathaniel Colby fund. –

Mr. Edwards stated that the last entry in the cash book is August 1875 – the Colby transaction 1881 to 4 does not appear –

Col. Leathes stated that the George Fowler mentioned above is a confirmed lunatic.

[8]   The thanks of the meeting were given to Mr Edwards for calling them together and for his information –

Mr James Peto proposed that Mr Fred Morse be appointed treasurer to the Suffolk Humane Society in the place of Mr Reeve deceased.

Lt.Col. Leathes seconded his nomination which was carried unanimously.

It was resolved unanimously that the chairman of this meeting do report the result of the same to the exors. of the late treasurer Mr Reeve and request them to deliver to Mr Morse all the books papers & documents relating to the Suffolk Humane Society. – Thanks to the chairman closed the meeting.

James Peto

[9] A letter of which the following is a copy was subsequently placed in the hands of Mr Morse. –

Carlton Colville
July 4th 1890

To C.J. Turner & C.W. Willett Esqres executors of the late R.H. Reeve Esq.

Gentlemen,

In response to the circular signed by Mr George Edwards, of which I enclose a copy – a meeting of such of the surviving members of the Suffolk Humane

Society as could be discovered was held at Mr Ellen's office on Wednesday the 2d inst. – present

Revd W.H. Andrews in the chair

James Peto Esq. J.P. Lt-Col. Leathes J.P. George Edwards Esq. J.P. J.L. Clemence Esq. Fred Morse Esq. and Mr T.E. Thirtle

[10] It was proposed by Mr James Peto seconded by Col. Leathes and carried unanimously that Mr Fred Morse be appointed treasurer to the above Society in the place of Mr R.H. Reeve deceased.

It was further resolved that I as chairman of this meeting should request that you should kindly deliver up to the said Fred Morse all the cash, securities for money, bank books, cash books, minute books, papers and correspondence relating to the Suffolk Humane Society presumed to be now in your possession as the exors. of the said R.H. Reeve.

Yours obediently,
W.H. Andrews
Chairman

[11]   Notice of Meeting

Lowestoft
7th January 1891

Dear Sir

Being appointed treasurer to the Suffolk Humane Society, in the place of the late R.H. Reeve Esq., at the meeting of July the 2nd last, I have, after much trouble, succeeded in obtaining possession of the derelict funds of your Society – I am therefore requested to ask you to meet the subscribers on Wednesday, the 14 day of January at half past two oclock at the Police Court, & to give me further instructions.

Yours very truly,
Fredk. Morse   honorary treasurer

[12] A meeting of the subscribers to the Suffolk Humane Society held at the Court House at half past two o'clock on the 14th of January 1891.

James Peto Esq.
in the chair

Present
George Edwards
Col. Leathes
Revd C. Steward
T. Thirtle
A. Stebbings
H.G. Woods
F. Morse
Revd Andrews
and the reporters of the
Lowestoft papers and Mr B. Preston

95

[13] The minutes of the last meeting were read and signed.
The treasurer then read the following report.

Gentlemen,

Since your last meeting held on the 2nd of July last, when you did me the honour to appoint me treasurer to your Society, I have made search for the funds remaining at your disposal, after the transfer of the lifeboat part of your business to the Royal National Lifeboat Institution as determined upon at the annual meeting of your Society on the 20th of **[14]** August 1873 –

As requested at the July meeting your chairman the Revd Mr Andrews wrote to Messrs Turner & Willett, the executors of Mr R.H. Reeve, your late secretary and treasurer (who died in 1888) requesting them to deliver to me all the books and papers relating to your Society (see copy back).

I regret however to have to state that I have only received a portion of them; which has rendered my search a matter of extreme difficulty, and even now prevents me producing such direct evidence of the position of affairs as I could wish – **[15]** I find the statement made at your last meeting by Mr Edwards to be substantially correct; and that there are funds at your disposal amounting to £720 7s. 9d. as I shall proceed to show –

In 1856 a Miss Crow left a legacy to the S.H.S. of £200 – this (A) is now invested in £214 14s. 8d. £2 15s. Consols in the names of R.H. Reeve, H.M. Leathes and G.G. Fowler – Messrs Gurney & Co. have received the dividends as they accrued, placing them to the current account of your Society – However since the death of Mr Reeve these dividends have not been received by Messsrs Gurney & Co. – A multiplicity of correspondence **[16]** with Messrs Gurney, the Bank of England, and the custodians of Mr Fowler became necessary causing much delay, and has only now been terminated to the satisfaction of the Bank of England by their paying up the overdue dividends to the 30th of December as appears in Messrs Gurneys account (B).

The cash balance at Messrs Gurneys is now £214 19s. 11d. which has principally arisen from the accumulated dividends from Miss Crow's legacy; but I must not omit to state also from the sum of £116 5s. 3d. the remainder of the Nathaniel Colby fund **[17]** after the death and burial of Colby and his wife \in 1884/ by Mr Edwards (C) \and £25 – J. Pratt's legacy in 1874./

At a special meeting of the Society held on the 1st of November 1820 a new fund was raised amongst the subscribers for the relief of Shipwrecked and maimed (and was lost sight of until discovered by Mr Edwards in 1856 when searching the books for another object) to be deposited in the Lowestoft Savings Bank – The last deposit book, being one of the books in the possession of your late secretary which has not yet been found; application was made to the Bank trustees and Mr Allerton has kindly furnished **[18]** me with another, carrying on the account from an old book up to last November the 20th including interest to that date amounting to £290 13s. 2d. (D).

Thus there remains to the credit of the Society

£2 15s. Consols	214 14 8
Deposit in Savings Bank	290 13 2
Cash at Gurneys & Co.	214 19 11
Showing a capital of	£ 720 7 9

<This amount would provide an annual income of £21 18s. 2d. at 3%>

Being in possession of the last cash book I can state that the last entry is at the date of August 1875.

[19] The last minute book is not yet found, I am therefore unable to give any information as to the cash proceedings of the Society; but it appears by Messrs Gurneys pass book to have been the giving of some small rewards for saving life, or the attempt to do so. –

I am Gentlemen

Your obedient servant

Fredk. Morse

Honorary Treasurer

(A) Copy of certificate from the Bank of England signed by S.O. Gray Accountant Genl.

(B) Society's pass book with Gurneys & Co.

(C) R.H. Reeve's receipt to Mr Edwards

(D) Society's passbook in the Lowestoft Savings Bank

January 14th 1891

[20]

The chairman asked what reply had been received to his letter of the 14th of July last – the treasurer read as follows –

122 High Street

Lowestoft

18th August 1890

Dear Sir,

R.H. Reeve deceased

Suffolk Humane Society

I send you herewith books and papers mentioned on other side – kindly acknowledge receipt.

Yours truly,

C.T. Turner

F. Morse Esq.

Lowestoft.

[21]

Suffolk Humane Society

List of Books papers &c.

Minute Book No 2   From 1806 to 1859
Treasurers account   From 1842 to 1869
          do   From 1869 to 1873
Bankers pass book
Stock receipts 3 & 2¾% for £214 14s. 8d.
Sundry bank cheques & correspondence

The treasurer reported that since this date he had received the Minute Book No. 1 from Mr B. Preston.

Mr Edwards made a statement and read the accompanying affidavit:

[22] Mr Preston Honorary Secretary of the Royal National Life Boat Institution attended and read the following resolution from the minute book of the Institution dated August 1873 –

"That the Suffolk Humane Society continue to carry out the objects for which it was originally formed, and for this purpose use the interest accruing from a certain fund raised in 1820 and from Miss Crow's legacy and from such subscriptions donations and legacies as may from time to time be given therefore."

This being a satisfactory confirmation of the correctness of Mr [23] Edwards affidavit embodying the fourth resolution of the meeting of the Society on the 20th of August 1873.

It was proposed by Mr Peto, seconded by Mr Stebbings and carried unanimously
1st   "That it is the opinion of this meeting that the arrangements made at the meeting of August 20th 1873 for utilizing the remaining funds of the Suffolk Humane Society have not been satisfactory and that it is desireable that some other arrangements be now made."

[24] It was proposed by Mr Thirtle seconded by Col. Leathes and carried
2nd   "That the resolution of this Society made at the general annual meeting of the 20th of August 1860 for the investment of the proceeds of Miss Crowe's legacy in Consols be and is hereby rescinded – That the 4th resolution of the meeting of this Society on the 20th of August 1873 namely

"That the Suffolk Humane Society continue to carry out the objects for which it was originally formed, and for this purpose use the interest accruing from a certain fund [25] raised in 1820, from Miss Crowe's Legacy and such subscriptions donations & legacies as may from time to time be given therefore" be and is hereby rescinded.

It was proposed by the Revd Steward seconded by Mr Stebbings and carried
3rd. That Colonel Leathes being now the sole trustee in whose name £214 14s. 8d. Consols stands, be requested to sell out the same

And that the treasurer withdraw the deposit of £290 13s. 2d. from the Lowestoft Savings Bank and that the products be placed to the [26] account of this Society at Messrs Gurneys & Co.

The thanks of this meeting to Mr Edwards for the trouble he had taken and to the chairman terminated the business.

James Peto

[*Printed notice inserted between pages* **26** *and* **27**]

Suffolk Humane Society.

LOWESTOFT,
6th November, 1891

DEAR SIR,

After much regrettable delay, your instructions have been carried out, and the funds of the above Society have been placed with Messrs Gurneys' & Co. I therefore request you to attend a meeting of the subscribers, on Tuesday, the 17th of

98

November, at Half-past Two o'clock in the afternoon, at the Police Court, to give me further instructions as to the disposal of the same.

Yours very truly,
FREDK. MORSE,
Treasurer.

[27]
List of names who had circulars posted to them

Andrews Revd W.H.
Birkbeck H
Birkbeck William
Colman J.J.
Clemence J.L.
Davey Miss
Devereux J.S.
Edwards G.
Fiske E.B.
Gurneys & Co.
Howett J.
Henderson J.
Leathes Col. H.M.
Morse F.
Peto J.
Steward Revd C.
Stebbings A.
Thirtle T.
Waddington H.J.
Wingfield Mrs.
Woodthorpe William
Woods H.G.
Woods J.K.   Youngman W.

[28] A meeting of the subscribers to the Suffolk Humane Society held at the Court House on Tuesday the 17th of November 1891 at half past two o'clock in the afternoon.

James Peto Esq. in the chair

Present
George Edwards
Col. H.M. Leathes
W. Woodthorpe
H.J. Waddington
Revd Chas. Steward
T. Thirtle
F. Morse

[29] The minutes of the meeting held on the 14ᵗʰ of January were read and signed. The Treasurer reported that shortly after the meeting of the 14ᵗʰ of January Col. Leathes instructed Messsrs Reeve & Mayhew (in accordance with the resolution passed at that meeting) to take the necessary steps to sell out the £214 14s. 8d. Consols – To carry this out Messrs Reeve and Mayhew had to send all the minute books and accounts belonging to the Society up to the Commissioners of Lunacy and with them lies all this unnecessary delay – It was not until **[30]** the 11th of this month that Messrs Reeve and Mayhew paid into Messrs Gurneys Bank the proceeds of the £214 14s. 8d. being £202 14s. 8d. – The amount withdrawn with interest £300 7s. 8d. from savings bank and £220 15s. 3d. at Messrs Gurneys & Co. makes £723 17s. 7d. the total amount the subscribers have to deal with less Messrs Reeve and Mayhew charges and treasurer's expences for printing &c.

The treasurer read some extracts from the old minute books relating to the subscriber's right to vote.

The treasurer read a letter written from the Sailors' Home dated February 1891 and signed   William Johnson
                                 Charles Tilmouth
                                 W.D. Sims

[31]   Mr Edwards considered that the meeting was in the unusual position of being called upon to act as executors in the distribution of the funds left by the society without any directions from the testator – But he would show from the records of the Society that it is not difficult to ascertain what were the wishes of the donors – and as honest men it would be their duty to distribute them as near as possible according to their intentions. –

The Society was originally a copy of the Royal Humane Society, but soon undertook the lifeboat work, devoting only a small part of their income to bestowing rewards. –

An entirely separate fund was originated in 1820 called the Shipwrecked and Maimed of which there now remains £300 7s. 8d. – The intentions of the subscribers to this and the work done so exactly **[32]** coincides with the work of the Lowestoft Sailors' Home that there cannot be two opinions as to the justice of turning this sum over to that local Institution. –

The expression "maimed" seems to infer the giving of medical assistance, this may well be provided for by a donation to that usefull Institution The Lowestoft Hospital.

In 1856 a Miss Crow left a legacy of £200 to the Lowestoft Life Boat Society, but altho' this was invested in Consols in the name of trustees no dividends were collected for thirteen years, nor can any employment of the money in accordance with her intentions be traced up to the present time. – you have learned from the treasurer that after much delay the Consols have been sold realizing £202 14s. 8d. and with interest accumulating   107 8s. 3d.
      since 1878 makes   £310 2s. 11d.
for which the society is indebted
      to Miss Crow –

[33]   We have no lifeboats now or machinery for giving rewards – But there are two Royal Societies still doing the work which we formerly did, and it would appear to be a simple act of justice to divide this sum between them. – It appears by the books that the sum expended by the Society in giving rewards was small compared

100

with that expended upon the boats, a fact which should be considered in any such division. –

By examination of such documents as are available it would appear that latterly the Society did not expend upon rewards more than on an average £5 a year, capitalizing this sum I arrive at £100 which would greatly assist the Royal Humane Society in doing what we have formerly done, a work which I learn from their secretary they are willing to do. –

[34]   This might leave some £200 to the Royal National Lifeboat Institution, but that Institution is very rich and lavish in expenditure, and I submit to you whether that sum given also to the Lowestoft Sailors' Home would not sufficiently meet the requirements of Miss Crow's legacy in favour of the shipwrecked mariner. –

I hold the receipt of the late treasurer, Mr Reeve, for £116 the remains of the sum collected by the Revd Mr Price and myself for the support of Nath. Colby, this was loyally returned as per special agreement for the use of the Society on such cases as that of Colbys, and I claim on behalf of the subscribers to that fund a voice in its disposal, & we desire that that sum should also be given to the Lowestoft Sailors' Home. –

[35]   I therefore now propose the following resolution to this meeting –

A conversation here took place as to the payment of the expences incurred in selling out the Consols –

When it was moved by Col. Leathes seconded by the Revd C. Steward & carried

"That the treasurer be authorized to pay the legal expences incurred in obtaining from the trustees Miss Crow's legacy fund, and also the expences for printing &c. since his appointment as treasurer". –

It was then proposed by Mr Edwards and seconded by Mr Woodthorpe

"That the sum of £100 be paid by the treasurer as a donation to the Royal Humane Society instituted in 1774 (4 Trafalgar Sq., Charing Cross, London) [36] with a letter shortly explaining the circumstances causing the gift. That the sum of £50 be paid by him as a donation to the Lowestoft Hospital, That after paying the above amounts the entire balance remaining in his hands be paid by him to the treasurer (Mr W. Johnson) of "The Lowestoft Sailors and Fishermans Home" for the uses and purposes of that Institution".

Colonel Leathes here rose & apologised to the meeting for being obliged to leave. He was however asked if he had any objections to the resolution proposed by Mr Edwards, and replied that he saw none. –

Mr Thirtle proposed an amendment to the resolution of Mr Edwards "That £100 be given to the Royal Humane Society, [37] £50 to the Lowestoft Hospital and £100 to the Widows and Orphans permanent fund established at Lowestoft in 1879 and the entire balance remaining to the Sailors Home".

This was seconded by Mr Morse & put to the meeting, when the following gentlemen voted for it

     Mr T.E. Thirtle
     Mr Fred. Morse
     Revd C. Steward and
     Mr Waddington

The chairman, Mr Woodthorp & Mr Edwards voting against it.

The amendment was therefore carried. It was proposed by Mr Edwards, seconded by Mr Morse & carried "That upon the completion of the above arrangement [38] the treasurer do prepare a statement of his accounts & submit the same to Col. Seppings who is requested kindly to audit the same, that thereupon the treasurer do call a meeting of the subscribers to inspect the same, and give any final instructions that may be necessary".

The thanks of the meeting were accorded to Mr Edwards and to the treasurer for the trouble they had taken, & to the chairman, this terminated the business.

[39 *blank]*

[40] The following letter <was sent to the Royal Humane Society> was written

Lowestoft 22 Dcr. 1891

Dear Sir

At a meeting held on the 17th <Decr> of November of the surviving members of the Suffolk Humane Society established in 1806 in imitation of your older Institution, I was directed to forward to you the enclosed cheque for £100 – as a donation to your Royal Humane Society established in 1774 –

After parting with our lifeboat establishment some small balances have been found and with accumulations of interest unused for several years. It has now been determined **[41]** to distribute this money amongst institutions most nearly coinciding with the object of the donors. – In presenting this sum to your society it is proper that cases of meritorious exertions for saving life, not uncommon on this dangerous coast, with its harbour and rivers, may meet with that attention from your society they may deserve.

Please be so good as to forward me a receipt from your treasurer for the above sum.

Yours obediently,
Fredk. Morse
Treasurer to the Suffolk Humane Society

Capt. J.H. Horne
      Secretary to the
Royal Humane Society
4 Trafalgar Square
London WC

[42]   Lowestoft 22 December 91

Dear Sir

The Suffolk Humane Society was established in 1806 – In 1820 another fund was raised in connection with the Society for the Shipwrecked and Maimed and in 1824[10] the first lifeboat placed on this coast was also the work of this Society –

But since parting with our lifeboat establishment the <remaining business of the> Society was allowed to drop off and the balances remaining were forgotten –

---

[10] This date is incorrect; the *Frances Ann* was put into service in 1807.

102

Through the energy of Mr George Edwards these sums with the accumulations of interest unused for several years have been found up.

[43]   It has now been determined to distribute this money amongst Institutions most nearly coinciding with the objects of the donors –

At a meeting of the surviving members of the Society held on the 17th of [December *deleted*] November I was directed to forward to the Hony. Secretary of the Lowestoft Hospital the sum of Fifty Pounds to be applied in furtherance of that Institution.

I herewith enclose the cheque to you and I shall feel obliged by your receipt for the above sum.

I remain dear Sir

F. Peskett Esq.   Yours truly

Hony. Secretary   Fredk. Morse

Lowestoft Hospital   Treasurer Suffolk Humane Society

[44]   Lowestoft Dcr. 22 / 91

Dear Sirs

The Suffolk Humane Society was established in 1806 – In 1820 another fund was raised in connection with the Society for the Shipwrecked and Maimed, and in 1824 the first lifeboat placed on this coast was also the work of this Society. But since parting with our lifeboat establishment the Society was allowed to drop off and the balances remaining in hand were forgotten –

Through the energy of Mr George Edwards these sums with the **[45]** accumulations of interest unused for several years have been found up – It has now been determined to distribute this money amongst Institutions most nearly coinciding with the objects of the donors.

At a meeting of the surviving members of the Society held on the 17th of <December> November I was directed to forward One hundred Pounds to be applied in furtherance of the Lowestoft Fishermens Widows and Orphans Permanent Fund – Herewith I enclose the cheque and I shall feel obliged to you as trustees of the beforenamed fund by your **[46]** returning to me your receipt for the above sum.

I remain dear Sirs

Yours truly

Fredk. Morse

Treasurer

Suffolk Humane Society

To

W. Youngman Esq.

B. Preston Esq.

Trustees of the Lowestoft Fishermens

Widows and Orphan Fund

[47]   Lowestoft
                                    March 29th 1892

Dear Sir

The Suffolk Humane Society was established in 1806 – In 1820 another fund was raised in connection with the Society for the Shipwrecked and Maimed and in 1824 the first lifeboat placed on this coast was also the work of this Society – The minute books alone attest the amount of admirable and philanthropic work carried on for some seventy years – But since parting with our lifeboat establishment the Society was allowed to drop **[48]** off and the balances remaining in hand were forgotten –

Through the energy of Mr George Edwards these sums with the accumulations of interest unused for several years have been found up –

It has now been determined to distribute this money amongst Institutions most nearly coinciding with the objects of the donors –

At a meeting of the surviving members of the Society held on the 17th <Decr> November it was resolved that after my paying certain specified sums to other Institutions, <then> the entire balance **[49]** remaining in my hands I was directed to pay over to the Treasurer of the Lowestoft Sailors and Fishermans Home "for the uses and purposes of that Institution." I now have the pleasure of enclosing you a cheque for £431 19s. 10d. and I shall feel obliged by your forwarding me your receipt for that amount.

                         I am dear Sir
                         Yours very truly
                         Fredk. Morse
Mr. Wm. Johnson   Treasurer Suffolk Humane
        Treasurer of   Society
The Lowestoft Sailors and Fishermans Home

Lowestoft Journal 16 April 1892
The last meeting of the surviving members of the old Suffolk Humane Society was held at Lowestoft Town Hall on April 12. The previous meeting had been on November 17, 1891.

£100 went to the Royal Humane Society, £50 to the Lowestoft Hospital, and £100 to the Lowestoft Widows' and Orphans' Permanent Funds, and the balance of £431 19s. 10d. to the Lowestoft Fishermen's and Sailors' Home. The books and documents of the Society were put into a tin box and handed over to the trustees of the Lowestoft Sailors' and Fishermen's Home.

BIBLIOGRAPHY

Cooper, E.R., *Storm Warriors of the Suffolk Coast* (London, 1937)

Dawson, A.J., *Britain's Life-Boats: The Story of a Century of Heroic Service* (London, 1923)

Gillingwater, Edmund, *An Historical Account of the Ancient Town of Lowestoft* (London, 1790)

Higgins, David, *The Beachmen: The Story of the Salvagers of the East Anglian Coast* (Lavenham, 1987)

Hunt, B.P.W. Stather, *Pakefield: The Church and the Village* (Lowestoft, 5th edition, 1938)

James, A., 'Frances Williams and the Anglesey Association for the Preservation of Life from Shipwreck, 1827–1857', *Anglesey Antiquarian Society and Field Club Transactions* (Caernavon, 1957), pp.20–25

Lewis, Richard, *History of the Life-Boat and its Work* (London, 1874)

Lukin, Lionel, *Description of an Unimmergible Life Boat* (London, 1820)

Malster, R., *Saved from the Sea* (Lavenham, 1974)

Malster, R., 'Suffolk Lifeboats – The First Quarter-Century', *The Mariner's Mirror* vol. 55 (1969), pp.263–80

Mitchley, J., *The Story of Lowestoft Lifeboats, Part 1: 1801–1876* (Lowestoft, 1974)

Moffat, Hugh, *East Anglia's First Railways* (Lavenham, 1987)

Osler, A.G., *Mr. Greathead's Lifeboats* (Newcastle upon Tyne, 1990)

Report of the Committee appointed to examine the Life-Boat Models submitted to Compete for the Premium offered by His Grace The Duke of Northumberland (London, 1851)

Robus, Frederick, *Lionel Lukin, of Dunmow, the Inventor of the Lifeboat* (Dunmow, 1925)

Warner, Oliver, *The Life-boat Service* (London, 1974)

White, William, *History, Gazetteer and Directory of Suffolk* (Sheffield, 1844; reprinted Newton Abbot, 1970)

INDEX OF PEOPLE AND PLACES

INDEX OF SUBJECTS

NAMES OF SHIPS AND BOATS

Aimwell, 50
Ann (brig of London), 24
Ann (brig of Sunderland), xliii–xlv, 24

Betsy, 16
Bishop Blaize, 24
Brittania (sloop of Ipswich), 15
Bywell (brig of Newcastle), xlvi–xlvii, 26, 77

Cammilla, 46
Catherine (of Sunderland), xxxvi–xxxvii
Costenside (brig of Newcastle), 24

David Ricardo (of London), 19–20, 42, 77
Defiance (smack of Plymouth), 15
Derwent (transport ship), xxiii
Don (brig of Whitby), 42
Dorset (sloop of London), xliii–xliv, 24
Dronningen (Norwegian brig), 77

Elizabeth (brig), 21
Elizabeth Henrietta (hoy of Papenburgh),
    xxxviii–xxxix
Endymion (barque), 54, 77

Fairy, HMS, 32
Farnacres (of Sunderland), 43
Fawn (brig of Sunderland), 24
Frances Ann (lifeboat), iv, xii, xxxvi, xli, xlv,
    xlvii, xlviii, 53
    building of, xxxi
    trials of, xxxiii, xlii, 12–13
Friends (brig of Scarborough), 17
Friendship, 52

George (brig of London), iv, xli, 24, 77
George and Henry (of Stockton), 28
Glenham Castle, 51
Good Intent (brig of Newcastle), 29
Gratitude (brig of Sunderland), 16

Harriett and John (brig of Sunderland), xliii–xliv
Hearts of Oak, 52
Helena (of Sunderland), 44–5

Isabella and Margaret, 77

James and Margaret (of Newcastle), 16
Jeanie (sloop of Hull), xxxix–xli

Jubilee (of Plymouth), 26

Kentish Lass, 65

Latona (barque), 59
Louisa (of Poole), 19
Lucks All, *see* Suck's All

Maria (sloop of Ipswich), 14
Meum et Tuum (fishing boat), 25n

Orwell, 83, 84
Oswy (brig of Shoreham), 83, 84

Peace (transport ship), xxiii
Pere Jolet (French vessel), 55
Prima Donna (fishing boat of Kessingland), 59
Prince Frederick (steamer), 19
Prince of Brazil (brig), 24, 77

Rebecca (of Bridlington), 16
Reliance (fishing boat), 62
Rescue, The, (Pakefield yawl), 31
Rochester (brig), xlv
Rose and Elizabeth, 28
Ross (brig of Pillhead), 17

Salem, 52
Sarah (sloop), 75
Sarah & Caroline (sloop of Woodbridge), iv, xli,
    24, 77
Sea (of Stockton or Hartlepool), 21, 29
Second Adventure (of Lowestoft), 60
Seaflower (yawl), xlvii, 27
Shamrock (steamer of Dublin), xliv
Sir Walter Scott (Leith smack), xlvii, 27–8
Snipe (gunbrig), HMS, xxx
    (brig of Faversham), 65
Speedwell (brig of Shields), 20, 77
Suck's All (schooner of London), 24, 77
Susan (schooner of Plymouth), 51

Tay (brig of Bridlington), 24
Tennant (brig of Stockton), 78–9
Thomas & Mary (brig of Newcastle), 24
Thomas Oliver (brig of Sunderland), 40
Trinity (Lowestoft yawl), xliii–xliv
True Briton (brig of Sunderland), 43

117